THE HEALTH OF

THE HEALTH OF NATIONS

Political Morality for the 21st Century

ROBERT VAN DE WEYER

GREEN BOOKS

First published in 1991 by
Green Books
Ford House, Hartland
Bideford, Devon EX39 6EE

Typeset in 12 on 14 point Erhardt
by Fine Line Publishing Services, Witney, Oxon

Printed by Biddles Ltd,
Guildford, Surrey
on recycled paper

British Library Catologuing in Publication Data
Van de Weyer, Robert 1950–
The health of nations : political morality for the 21st Century.
1. Politics. Ethics
I. Title
172

ISBN 1–870098–39–0

Contents

Introduction

Just over two centuries ago, in 1776, an obscure Scottish scholar published a book entitled *An Inquiry into the Nature and Causes of the Wealth of Nations*—usually abbreviated as *The Wealth of Nations*. It is without doubt the most influential work of social philosophy ever written. Within fifteen years it was being quoted in the House of Commons to support the abolition of restrictions on international trade. And in the decades after the defeat of Napoleon in 1815 it was the Bible for economic and social policy, first for the British government and then for governments throughout Europe and North America. Adam Smith foresaw the explosion of economic and scientific energy which we now call the Industrial Revolution, and he described the subtle mix of policies which the state must enact to release that energy. The political and economic system which we describe as 'capitalism' or 'the free market' is to a great extent the brainchild of Adam Smith. Today that child has come to birth in every corner of the globe, and even now is transforming the lives of the remotest tribespeople. And though few have heard his name, the attitudes and actions of all of us continue to be moulded by the thoughts of this shy prophet.

Now, however, the age of Smith must end—either that or the human race will destroy itself. Previous civilisations have caused great damage to the natural environment. But the ecological destruction wrought by industrial capitalism is beyond the worst nightmares of all previous eras. We do not know precisely the speed at which we are heading for catastrophe, but few now doubt that if our economic and social system continues unchanged, the coming millennium will see the extinction of our species. It is a matter of the utmost urgency that we discover a moral philosophy and a political system that can steer our civilisation on to a fresh course—towards an era when our species can live in harmony with the other fauna and flora with which we share this fragile planet.

This book is an attempt to outline such a philosophy and such a system. None of the elements within the book is original—but nor were the elements that formed *The Wealth of Nations*. If this book possesses originality, it is in connecting disparate elements to form a new compound. And it is purposely brief and simple. The power of Smith's ideas lay in their elegant simplicity, which enabled them to be expressed even in children's fables. The salvation of our species must also lie in simple ideas within the grasp of children.

To Adam Smith the attainment of wealth was the primary aim of both the individual and the nation. Such an attitude seemed to many at the time quite shocking, since power and status had long been regarded as the proper objectives of people and nations. Today *health* must become our aim: the health of individuals and society, which is inextricably bound up with ecological health. This does not imply some dreary material poverty, in which culture and living standards are sacrificed in the name of some greater physical, environmental and spiritual good. On the contrary, a healthy civilisation is rich, both culturally and economically, because Nature herself is rich. If this book has one overriding purpose, it is to show that a society which is healthy in body and spirit will also be enjoyable.

Since our concern is with human civilisation as a whole, I have used very broad brush-strokes, showing the connections between the various aspects of civilisation, while leaving the details blurred. I am bound to raise questions in readers' minds which are left unanswered.

In the end, however, there is only one question which matters. Can the picture which I try to paint of a humane and sustainable civilisation be realised in practice? Adam Smith in his day was accused of wishful, utopian thinking; later generations have more often regarded him as a dismal realist. As someone trained in the discipline which Smith himself fathered, economics, I have attempted to emulate his realism. But my own response to those who accuse me of utopianism is to turn the accusation on its head. Do any of us imagine that our present pattern of life can last much longer? Those who act as if it can are guilty of the most ludicrous self-delusion. Like it or not, the future will be vastly different from the present. Let our imagination be bold, even foolhardy, as we look towards the coming millennium—that our descendants may have cause to bless us for our generous vision, not curse us for selfish myopia.

PART ONE

DISEASE

I

Divided Universe

Science Without God

We live in a very peculiar culture. Every other culture in history, and every non-Western culture across the world, has acknowledged that human destiny is largely outside human control. Some cultures have regarded the wind and the rain, the sun and the moon as divine, giving names to the gods which preside over them. Others have believed in a single god whose almighty power gives life to every creature, from the elephant to the ant, and whose invisible hand pushes every cloud across the sky and lights every star. Some have simply stared in awe at the great mystery of Nature, not daring to speculate about supernatural forces.

And every culture, apart from our own, has warned against trying to defy or supersede the laws of Nature. To the ancient Hebrew tribes of Israel, described in the Old Testament, human sin consisted, purely and simply, in humanity trying to usurp the place of God. The temptation, to which Adam and Eve succumbed when they ate the forbidden fruit, was to decide for themselves between good and evil—to ignore God's laws, embodied

13

in his creation, and to live by their own power and wit. As Augustine put it, reflecting on the story of Genesis, the root of sin is pride. And in acknowledging human sin, the Hebrew tribes sought to resist it, their prophets urging them to submit to God's laws. The Old Testament is the history of tyrants and demagogues striving to control the world around them, to satisfy their lust for power, and yet ultimately being defeated by the superior forces of Nature—and thus of God. Every tribe and every culture has similar myths and similar tales with the same lessons.

The story of Jesus in the wilderness re-enacts the temptation of Adam and Eve, making even more explicit the source of sin; the difference is that Jesus overcomes the devil. In the first temptation Jesus is offered control of Nature herself, by turning stones into bread. In the second he is offered power over people, by becoming emperor of the world. And in the third he is offered power over Nature and people together: by being held up by angels as he jumped from the Jerusalem temple, he would both defy the laws of gravity, and dazzle his spectators into accepting his authority. Again one can find similar stories involving Krishna, the Buddha and Mohammed, all with the same double message: that the natural order is divine, to be treated with respect and reverence by the human race; and that those who seek to overturn the laws of Nature will themselves be destroyed.

Modern Western civilisation is the first culture in history to reject that message. Certainly there have been numerous kings and emperors who have imagined themselves as gods, even claiming mastery over the wind and the rain; but their arrogance has been in defiance of the religion and philosophy of their culture. The philosophy and the way of life that has developed in Europe over recent centuries, and spread outwards across the world, exalts such arrogance as both righteous and sane. God has been banished from his creation, and is now largely forgotten—except at times of sickness and death when our grip on Nature momentarily seems to weaken.

The philosophers who first clearly pronounced God's banishment in the seventeenth and eighteenth centuries became known as deists. They acknowledged that God had created the universe, but they compared him with a clockmaker who designed and made the perfect clock, with the parts moving in absolute harmony, and then withdrew to admire from afar his beautiful creation. The universe was thus portrayed as a vast machine, whose mysteries could be penetrated by patient investigation. There could be no miracles, no supernatural forces at work, and prayer had no purpose apart perhaps from offering solace to those who prayed. Every event must have a natural cause which human beings could discover if they were intelligent enough. It is small wonder that by the nineteenth century God was disappearing altogether from the philosophical scene. In Laplace's famous words to Napoleon: 'God is a hypothesis which is quite unnecessary.'

The philosophers were not, of course, thinking in an intellectual vacuum. On the contrary, they were responding to an intellectual hurricane that over the previous four centuries had blown with increasing ferocity across the face of the globe.

Before the sixteenth century, theologians and laymen alike believed that the essential truths about the universe were revealed by God, primarily through the Scriptures; human reason could simply codify and clarify those truths. But the first scientists, such as Copernicus and Galileo, looked upwards to the sky not to spy heaven, but to observe the movements of stars and planets, and then check whether the Bible's picture of the universe was accurate. It is hardly surprising that when they suggested that the earth was not the centre of the universe, but merely one small planet revolving round the sun, they were condemned as heretics. Not only did their ideas fly in the face of Scripture, but far worse, their method was in stark contradiction to the godly endeavour of the theologians.

Papal threats and punishments could not, however, suppress these first scientific stirrings. Moreover it was soon clear that science could not only investigate Nature, but also change her, adapting her laws to the benefit of humankind. One of the early proponents of the scientific method, Francis Bacon, believed that 'our most noble and healthy ambition ... is to extend the power and dominion of the human race over the universe'. To this end scientists must 'make careful observations of the particular workings of nature ... gradually seeking to discern general principles'. Then once the processes of cause and effect have been laid bare, the scientist can intervene to change the course of Nature. Bacon coined the adage which has guided this whole scientific venture: 'Nature to be commanded must be obeyed.'

At first science wore a cloak of humility. Bacon wrote of the scientist 'turning the pages of creation with awe', seeing there 'the stamp of the Creator'. And the greatest scientist of this period, Isaac Newton, spent as much time on theology as on science, struggling to reconcile his scientific work with his Christian faith. Yet Newton's picture of the universe persuaded many others, if not himself, that God now played no active part in his creation since, according to Newton, all the elements and objects of the universe moved in perfect order. Certainly God should be thanked for instituting that order; but hereafter God could safely be ignored. Newton himself seemed aware of the danger lurking in the heart of science, that if the scientist tampered too effectively with Nature to improve the human lot, the delicate balance of God's creation could be reduced to chaos. But, then as now, scientists were too busy mastering their own corner of creation to worry about the natural order as a whole. And the businessmen who harnessed the scientific discoveries, then as now, rarely allowed such anxieties to cloud their energetic optimism.

Science itself has, of course, changed greatly since those pioneering days. Newton's laws have been super-seded by—or rather subsumed within—Einstein's theories; and the scientists' command over Nature extends beyond the wildest dreams of Francis Bacon. Yet the attitudes and ambitions nurtured by Bacon and his contemporaries have passed down to us unaltered. We imbibe the scientific, deistic outlook with our mother's milk; and the myth of scientific omnipotence is as powerful within our imaginations as the story of Genesis was within the Hebrew mind.

This myth of scientific omnipotence has engendered a second, even more potent myth: that of progress. To our pre-scientific ancestors, time, at least as it manifested itself in Nature and in their material lives, was circular. There was the cycle of seasons, and thus the cycle of sowing and harvesting, repeated each year; and there was the cycle, shared by all living creatures, of birth and death. Over and above that there was a political cycle, of rulers coming and going. Our ancestors were profoundly aware of variations in the cycles: some years brought abundant harvests, others brought famine; and some rulers brought peace and justice while others brought conflict and terror. But these variations were themselves part of a divine, moral cycle, of good and evil, with no one imagining that one would permanently vanquish the other until the final day of judgement.

Science and technology have pulled this circle out into a straight line. Bacon's image of science as an empire, extending its dominion over the universe, promised that each generation would enjoy greater control over the natural order than the last. And this in turn held out the prospect of ever-increasing comfort and prosperity for the human race, as it turns the laws of Nature to its own advantage. Today we take Bacon's promise for granted. We expect to enjoy a higher standard of living than that of our parents. We assume that each new product from our factories is an advance on its pred-ecessor. Most important of all, our culture teaches us that change should always be for the better, never for the worse.

Bacon himself foresaw that progress would accelerate. Not only would one scientific discovery lead to another, but each door that opened into the mysterious 'citadel of Nature' would reveal many more doors through which to walk. And this too has become accepted as both inevitable and desirable. We enjoy looking back and enumerating the vast changes that have occurred between our grand-parents', or even our parents', time and our own. And we are delighted at the speed at which fashions change and products become obsolete. Our nagging dissatisfactions with our lot are assuaged by the prospect of improve-ment.

The myths of any culture point to its image of paradise; and our image is that of a city in which the triumph of science is complete. Our houses should be heated to the most comfortable temperature, regardless of the season. Our food should have both its taste and its nutrients adjusted to suit our palate and our health. Our muscles should be kept supple and strong in gymnasia where each piece of equipment has been designed to achieve maximum results in the minimum time. We should be carried from place to place in cars that replicate the comforts of home. Even in our gardens, where Nature is closest, the grass should be severely mown, and any plant not purchased from a shop should be destroyed as a weed.

We are, of course, acutely aware of the imperfections, even the grotesque misery, of some of the actual cities we have built. And, just as Newton awkwardly wondered whether Nature would one day be destroyed by men's conquest of her, so most of us occasionally wonder whether the paradise we are trying to build is perhaps a prison, whether in promising to control Nature for our benefit, science and technology are in fact controlling us. Spurred by these fears, increasing numbers throughout the Western world are retreating to the countryside hoping to find a different relationship with Nature. But, more often than not, the myths prove too powerful, the image of the urban paradise too strong. So we turn our villages into small cities, cutting back the hedgerows, mowing the wild grass, paving the verges, building houses no different from those we have left, and buying food in the same supermarkets from the same factories. The only difference, of course, is that our dependence on that supreme symbol of our civilisation, the motor car, becomes total.

The Growth Machine

Just as in our relationship with Nature we have raised pride from a sin to a virtue, so in our relationships with one another self-interest has enjoyed a similar elevation. And God has in the process been banished from social affairs as well. In fact, historically and philosophically, the two attitudes are two aspects of a single transformation in our culture.

Once people had grasped Newton's vision of the mechanical universe, they naturally wondered if the same perfect order could be found in social affairs. The equivalent to Newton's scientific laws of motion would be social laws of motivation. Thus people began to ask whether the desires and wants which motivate human conduct, if allowed to express themselves freely, would create a prosperous well-ordered society. By the early eighteenth century this had led to a fierce argument over the degree to which human beings are naturally benevolent. Some, like Hobbes, believed that humans are basically selfish and egotistical, and hence need strong social and political institutions, headed by an absolute ruler, to prevent chaos—to avoid a 'war of all against all'. Others, like the Earl of Shaftesbury, asserted that humans possess a natural compassion, through which each individual can feel the pains and pleasures of others, and that this compassion tempers self-interest, and hence maintains social order without the need of political intervention.

Both the religious and the political battle lines were highly confused in this debate. Shaftesbury and his allies imagined that they were defending Christianity against the atheistic attacks of Hobbes's supporters. Yet even some churchmen recognised that Christian morality had never depended on some vague appeal to natural compassion; on the contrary, its starting point is a robust recognition of humanity's natural selfishness—'original sin' in traditional terminology—and it therefore addresses itself to the means by which the human heart can be changed.

Thus Hobbes's view of human motivation seemed more firmly rooted in reality, at least as regards economic activity. But Hobbes did not hold out any promise of a perfect social order, in which unfettered human action would lead to the greatest good for all; on the contrary, to Hobbes individuals left to themselves would revert to savages—man's 'natural' state in which life is 'nasty, brutish and short'. The genius who discovered in human society the same apparent natural order that Newton had discerned in the universe was the self-effacing Scottish professor, Adam Smith, recognised as the father of modern economics. Few have read his masterpiece *The Wealth of Nations* which appeared in 1776, but the minds of all of us have been moulded by its contents. Smith agreed with Hobbes that self-interest is the mainspring of human action in society. A man goes to the baker, not to help the baker, but to satisfy his own hunger; and the baker sells him bread, not out of compassion for his empty stomach, but to earn money with which to satisfy his own needs. Only within the family and amongst friends do more charitable impulses apply.

Like Newton, Smith possessed sufficient native religion to want to acknowledge the glory of God's handiwork. Thus he spoke of an 'invisible hand' which used the self-interested actions of individuals to achieve the greatest prosperity for the nation as a whole. This invisible hand is not an active divine power, performing a miracle each time bread is bought and sold. Rather it is a kind of social law of motion, which guides human conduct regardless of whether people are aware of it or not. In the society which Smith was describing it could most clearly be seen in the operation of markets.

Smith analysed the market-place with such simplicity and clarity that his words were soon being quoted by politicians, and his theories were being distilled into fables to be read to children. Smith's political economy is now part of the mental furniture of the Western mind, taken for granted in our social outlook. In Smith's words, the consumer is sovereign. Consumers express their desires through what they purchase in the shops. Producers in turn will only be able to earn a living if they make and sell what consumers want. Thus the factories which are built and the work people do will precisely reflect the goods which consumers demand. If in addition there is vigorous competition, in which there are many firms in each market, producers will be compelled to use resources efficiently, or else go out of business; so the economy as a whole will maximise its output. The crux of this social law of motion is that everyone should buy what will give most satisfaction to himself or herself; producers should pursue the highest profits, since these will only be gained if they are meeting the consumers' wishes; and workers should seek the highest wages, since this will guide them to the most productive occupation.

It is an alluring vision. Original sin is turned into social virtue; greed becomes love. And today the disciples of the Scottish prophet are far more numerous and more widespread than those which Christ, Mohammed and the Buddha ever obtained—even though few know his name. Indeed, the ideas of Smith soon acquired the status of a secular religion. In Britain in the early nineteenth century, governesses to middle-class families were required to be familiar not only with the Bible but also with the rudiments of Smith's political economy. So small children were taught that, while Jesus Christ was the master of their private, spiritual lives, Adam Smith reigned supreme in public affairs. Later in the century, as the Christian mission to Africa and Asia gathered momentum, such influential churchmen as the explorer David Livingstone saw commerce and the gospel as equal partners: the natives should first be tempted with the material fruits of Western industry, in order to soften their hearts for the intangible benefits of Western religion. It is hardly surprising that Adam Smith—or rather his ideas—have proved more attractive than the teachings of Jesus. And today there is almost no corner of the globe that has not to some degree been caught in the web of the international market.

The triumph and the vindication of Smith's vision within Britain was the Industrial Revolution, which rapidly spread across Europe and North America. Smith himself saw the first stirrings of this revolution in the 1770s in the small water-powered factories of the Pennines and western Scotland. Hitherto production had taken place almost entirely in people's own cottages, with usually a single person working alone. The new breed of entrepreneur discovered that by employing a group of men in a large building, and by dividing the production process into a series of small, simple tasks, the output per worker could be multiplied ten- or twenty-fold. The advent of the steam-engine brought the factories down on to the plains, and soon hundreds of men were sweating for fourteen or sixteen hours a day in factories which dwarfed even cathedrals. In the half-century after the publication of *The Wealth of Nations* the face of Britain was transformed by the operation of the free market. Almost half the entire population left their villages to inhabit the sprawling industrial cities, and the economy enjoyed rates of growth unimagined in any previous era.

The families who left their ancestral villages were driven by the dire poverty of the countryside, so that even the meagre wages of the factories seemed preferable. But, while rural poverty was to a great extent hidden, the squalid conditions of the new industrial cities were terrifyingly visible. Moreover the concentration of poverty soon turned the apathetic misery of the peasant into a seething discontent and anger, easily exploited by revolutionary agitators. By the middle years of the nineteenth century a new creed had emerged—socialism, which seemed directly opposed to the free market faith of Smith. The socialists argued that the free-market created two economic classes, the capitalists who owned the factories, and the labourers who worked in them. The wealth of the capitalists would grow, as they ploughed profits back into their businesses, while the wages of the workers would remain stagnant. In Karl Marx's phrase, capitalist profit is the 'surplus value' from the workers' efforts, and hence amounts to legalised robbery. So the socialists proposed that the factories should be owned by the state, to ensure that workers enjoyed the full fruits of their labours.

Since the mid-nineteenth century the politics of Europe, and indeed of almost the entire world, has seemed to be polarised between 'right' and 'left', capitalist and socialist; or, more precisely, it has been pictured as a line running from right to left, with political parties and individuals locating themselves somewhere along this line. We have inherited a one-dimensional view of politics. Thus political debate has been reduced to petty squabbles about which point on the line is the most desirable. Throughout the nineteenth century capitalism remained dominant, although socialist ideas enjoyed growing appeal amongst both the working classes and intellectuals. In the middle years of the twentieth century, from the 1920s to the 1970s, socialism in varying forms has been popular, not only in Europe, but in much of Africa, Asia and South America too. From the 1980s, and especially since the collapse of Eastern Europe's socialist regimes in 1989, the pure ideology of Adam Smith has enjoyed a new lease of life.

Yet, despite the polarity of right and left, capitalism and socialism are in truth variations on a single theme. At heart both are judged—and judge themselves—by the same criterion: their ability to turn the resources of the earth into goods to be consumed. And they both measure this ability by the same crude yardstick: the rate of economic growth. In the heyday of Soviet five-year plans, as steel mills and research institutes proliferated across the Russian empire, it could be argued that socialism was more successful in harnessing Nature for humankind's benefit. Today the empty shelves in socialist shops, and the shoddiness of those goods which are manufactured in socialist factories, reassure capitalists that their system is more efficient.

The dispute between the two ideologies concerned the distribution of the fruits of economic growth. But here the differences are more apparent than real. There was never any intrinsic reason why capital—the factories and the machines within them—should all be owned by a single small class of people; nor was it ever inevitable that workers' wages should remain so low that a man could barely feed himself, let alone his family. Indeed some forms of socialism—such as that advocated by the co-operative movement—consisted in the workers becoming shareholders of their own factories, so that in effect the two classes of capitalist and worker would merge. Although co-operatives in their original form have not flourished, many of their aims have been realised. By the late nineteenth century wages throughout Europe and North America were rising rapidly; and, despite a hiccup during the 1920s, they have continued to increase. Moreover, today a majority of people own shares, either directly or, more commonly, indirectly through pension funds, unit trusts and the like. At the same time there are few people who are willing or able to live in idle luxury, enjoying the ill-gotten profits from factories far away; indeed, the idea of a 'leisure class' is frowned upon, and the offspring of even the richest families are expected to pursue careers. We are all capitalists and all workers now.

There remain, of course, grotesque inequalities in living standards, which in some respects are worse than those of the nineteenth century. The annual salary of a senior banker or the managing director of a large company is almost double the life-time wage of the person who cleans his or her office. The squalor in which the poorest 20 per cent of the population of Europe and North America live is only marginally less frightening than the conditions in Manchester or Glasgow under Queen Victoria. And, above all, there is the moral obscenity of hundreds of millions of people living on the edge of starvation in Africa and Asia while incomes continue to rise in Europe and America. But these are problems which socialism does not address; indeed they are insoluble within the one-dimensional thinking of left-right politics.

Capitalism and socialism are both ideologies of power. Their aim is power over the natural environment; and the means through which they achieve that aim is a power struggle between people. In a capitalist economy the power struggle occurs in the market-place, as firms compete for custom, each trying to expand its share of the market at the expense of its rivals. As Marx himself acknowledged, competition in the market-place has proved the most effective method ever devised for increasing production—and thus for enhancing humanity's control over Nature. To rid capitalism of its inequalities Marx and his comrades wanted to channel that competitive energy into the service of the state. Unfortunately where socialism has been vigorously practised, the power struggles of the market-place have simply been transferred to the offices of the state bureaucracy, to the detriment both of economic prosperity and of individual freedom. Those who have experienced socialism now generally prefer the jungle of the market to the combination of bureaucratic incompetence and state-sponsored brutality of a left-wing regime. By the criterion of success which capitalism and socialism both share, capitalism has emerged the winner. It is this criterion, however, that people are now starting to doubt.

2

Divided Self

Body Without Spirit

When we look outwards, we no longer see God in the rain and the sunshine, in the growing crops and the gambolling lambs. Nor, when we look inwards, do we see God in our thoughts and feelings, in our actions and reactions. It is not that we deny the existence of God; on the contrary, opinion polls consistently show a substantial majority believing in a divine being. It is rather that we exclude him both from Nature and from our own daily experience of life.

As we saw earlier, the division of God from Nature occurred with the rise of science in the seventeenth century, and was sealed by the advent of market economies in the eighteenth. The exclusion of God from ordinary personal experience has a far longer history, involving fierce debates about religion itself. It led to a view of human nature which, morally speaking, encourages the rise of science and market economies, and which in our time has led to very peculiar attitudes to human health and education.

31

Two and a half millennia ago the Greek philosopher Plato pictured the human being as a soul imprisoned within a body; and he asserted that the true object of our lives is to release the soul from all bodily attachments. Although the precise distinction between body and soul is sometimes blurred in the writings of Plato and his followers, it is clear that bodily attachments include our desire for food, warmth and sex, while the soul embraces our emotions of love and our appreciation of beauty. Plato urged his disciples to deprive themselves of all physical pleasures, so that the soul could devote itself to divine contemplation.

To the Hebrew tribespeople of the Old Testament, and to Jesus Christ and his apostles, such ideas would have seemed nonsensical. They took it for granted that body and soul are both created and hallowed by God. But as Christianity spread across the Mediterranean, it was adapted to appeal to people steeped in Platonic philosophy. Platonist sects arose within the Church whose adherents were taught to despise the body, and to treat the whole material realm with contempt. Most imitated the physical austerities of Plato, but some went to the other extreme, showing their hatred of the body by indulging in the most lavish orgies. Gnosticism was eventually condemned by the Church authorities, but its influence remained profound. The greatest theologian of the first millennium, Augustine of Hippo, spent his early years in such a sect; and even after he converted to orthodox Christianity, the rigid division between body and soul was the centrepiece of his theology. He saw the human condition as a deadly rivalry between the City of God, in which spiritual values are supreme, and the earthly city, in which greed and lust rule men's hearts.

Augustine, above all people, was responsible for the conviction, which is still held by some Christians, that sex is sinful. And in the early medieval period much of the Church's energies were absorbed in a moral vendetta against the extended family. The main underlying purpose was to maximise the chances of a man dying without legitimate male heir, so that his property would revert to the Church. Thus adoption was outlawed, the definition of incest was widely extended—at one stage including even sixth cousins—so that many partnerships became illegal; and the levirate marriage, through which a man must marry his brother's widow, was abolished, forcing many bereaved women to enter convents, taking their property with them. The Church's wealth grew rapidly, and in the process the tribal family system of Europe was destroyed. Christianity thus found itself obsessed with sexual immorality, coming to regard any form of sexual pleasure, even with a legitimate partner, as immoral. As Augustine had taught, it was within the sexual sphere that the division between body and soul was most severely applied.

The breakdown of the tribal family had a further, unforeseen consequence which widened the gulf still further. In a tribal society economic activity takes place within the context of close personal relationships; so the physical and emotional aspects of human life are unified. Once the family bond is broken, people are free to pursue economic gain without emotional constraint. It is this problem which came to absorb the theologians of the late medieval period.

From about the twelfth century onwards the market economy developed rapidly. For a further 600 years families continued to grow most of their own food, build their own homes, and make many of their own tools and utensils. But alongside this subsistence economy a new merchant class arose, trading in luxury goods such as fine textiles. A wool merchant would typically buy fleeces from a peasant farmer, taking them to a master weaver who would employ people to spin and weave the wool, and then the merchant would sell the finished cloth in one of the major market towns. Soon these merchants were accumulating wealth that was the envy of kings and barons. The theologians at first regarded this trade as evil, albeit a necessary evil, since it encouraged greed, and they urged the merchants to do penance—especially by donating generously to the Church. In addition they attempted to define the 'just price' for goods, based on the labour expended in its manufacture. They did, however, stand firm against usury—charging interest on loans—which they condemned as a mortal sin.

But by the fifteenth century the theologians found themselves in dignified retreat. As the merchants themselves were pointing out, in a free competitive market the price will automatically tend to be 'just', since any merchant who attempts to overcharge will be undercut by his rivals. At the same time a new breed of bankers, on whose loans much trade depended, argued that usury was really the 'just price' of money, since the lender should properly be compensated for losing the use of his money. As for penance, many merchants happily lavished their wealth on magnificent church buildings, motivated no doubt by a mixture of pride and piety.

But beneath these tortuous economic and ethical arguments, a much more important moral and spiritual point was being conceded by the theologians. In the tribal or feudal village, economic activity was guided by time-honoured customs and mutual obligations, of which the Church was regarded as the moral guardian: every aspect of life was under its moral oversight. The market economy, however, defied moral control, and the Church was compelled to accept that the buying and selling of goods—and thus the buying and selling of labour—was beyond its moral and spiritual jurisdiction. And as the market expanded, so the area of society under moral control shrank. The Church in effect applied a Platonic dualism to human activity: religion and morality covers only private family life, while economic, and thus political, affairs are beyond divine control.

This dualism was hallowed by the Protestant Reformation of the sixteenth century. Both Luther and Calvin looked back to Augustine for inspiration, eagerly adopting his image of two cities. To them the City of God was the invisible company of souls whom God had saved, and the Church was the visible embodiment of this company. The Church's sole task, therefore, was to save more souls and to nurture those already saved. In theory this could imply a moral indifference to all human behaviour, including even sexual activity; and indeed there were some groups, both Protestant and Catholic, who took this extreme position, regarding religion as a purely spiritual matter consisting solely of quiet contemplation. In practice, however, most believed that salvation should find expression in sexual propriety and private generosity. As for the earthly city, Luther and Calvin believed that human beings in their worldly affairs would inevitably act in a greedy and selfish way; thus the Church's only role was to uphold established authority, and thus prevent chaos. It is from this theology that has derived the extraordinary willingness of church leaders over the centuries to support even the worst tyrants, including Hitler.

The full moral and philosophical impact of the Church's retreat from the world only came to be felt in the seventeenth and eighteenth centuries. The kernel of every religion is that human nature can be transformed, that feelings and attitudes which are destructive and selfish can be made creative and loving. Since Christianity now confined this transformation to the narrowest arena of human experience, people soon came to regard the bulk of human feelings and attitudes as fixed and immutable. Philosophers who turned their attention to psychology portrayed human nature merely as a bundle of wants and desires, and they believed that all human action could be analysed as maximising pleasure and minimising pain. In the words of Jeremy Bentham, the human brain is a 'felicific calculator'; and any attempt to justify human conduct on moral rather than utilitarian grounds is mere hypocrisy. In this context the debate between Hobbes and Shaftesbury over whether human beings are capable of 'compassion' was really a minor squabble, since both accepted without question that human nature is fixed. It is a comparatively easy matter to include a degree of compassion in Bentham's felicific calculus, thus asserting that the rich man who throws a coin into the beggar's bowl is really only assuaging his own compassionate pain.

Philosophers are often led by their own logic to extreme conclusions, while less intelligent mortals hold back. So not surprisingly many people felt uneasy about reducing all life to a crude calculation of pleasure and pain. As in political economy, so in morality Adam Smith furnished the solution which to this day most of us accept. He followed the Church in dividing life into two distinct spheres, the public and the private. The public sphere includes our daily work in the factory and on the farm, and includes the buying and selling of goods. Here self- interest rules, tempered only by a desire for honest and fair dealing. Thus in deciding how to spend our income or what work to do our brains are indeed felicific calculators, judging everything according to our own pleasure and pain. The private sphere embraces personal relationships within the family and amongst friends. Here the guiding force is love, which may prompt action of self-sacrifice that cannot be reduced to any measurement of pleasure and pain. Smith thus pictured the human animal as inhabiting two quite separate moral orders, in which utterly different values apply; and he marvelled at the ease with which we can move from one sphere to the other, altering our attitudes entirely without a second thought.

It is indeed an astonishing portrayal of human life which, stated baldly, seems quite implausible. Yet it has a long intellectual ancestry, going back via Protestant and medieval theology to the philosophy of ancient Greece. And it is the moral picture which we today take for granted. We admire ambitious self-interest in the market-place, and we urge our young people to pursue wealth and status in their careers. On the other hand we applaud kindness and self-sacrifice within the family, praising those who set aside their own interest on behalf of their children or their friends.

The reason why we can so easily accept such moral schizophrenia is that it is embodied within society itself. When the Church waged war against the extended tribal family fifteen centuries ago, it began a social process that culminated in the Industrial Revolution. As the tribal system collapsed, to be replaced by the feudal village, it took economic activity out of the tight network of family loyalties, and transferred it to a looser network based on locality and on common allegiance to the lord of the manor. The merchants then absorbed a growing proportion of labour and resources into the vast, impersonal web of the market economy. The climax came when people left their villages altogether, to work in huge factories and live in sprawling cities.

The family thus became an isolated social entity, from which people go out to work and come back to rest.

Indeed in the early nineteenth century sheer poverty drove both men and women to earn wages, leaving even tiny babies to fend for themselves. By the late nineteenth century rising wages enabled most women to stay at home, rearing the children, cooking the meals, and even making the family clothes. But the era of the housewife lasted only a few decades, and now at the end of the twentieth century the majority of women throughout Europe and North America again go out to work. So even cooking is done largely in factories, while knitting, sewing and the other old household arts are practised less and less.

Thus the moral dualism of Adam Smith merely reflects the social dualism of our daily lives. If we try to stand back and look at ourselves from afar, we do seem very peculiar, apparently gripped by a moral insanity. But from inside it all feels perfectly natural. It is not surprising that most people continue to profess religious belief, and privately hold those beliefs dear, since our homes and families remain sacred. Yet it is equally unsurprising that institutional religion—going to church—is regarded as a harmless hobby practised by a small minority, bereft of any public purpose or role.

The Welfare Machine

Of all the social and economic institutions in modern Western society, the two that people will most vigorously defend are public education and public health care. We are passionately committed to the notion that the state

39

should provide free or subsidised schooling to our children and free or subsidised treatment to our sick. The extent of the subsidies varies from country to country, but all governments provide them. These are two services where the market economy must not intrude, and which we as individuals have no desire to control or manage; we want to be passive recipients of state benevolence.

On the face of it our attachment to state education and health care seems odd, inconsistent with our commitment to the free market. We believe, quite rightly, that the state would be a poor provider of bread or of television sets; the grotesque inefficiencies of state-run industries in Eastern Europe prior to the collapse of communism furnish ample evidence of the validity of this view. We are justifiably suspicious of state control of the mass media, fearful that even benevolent governments would try to manipulate popular attitudes. And we vigorously defend all attempts by the state to interfere in our private affairs. Yet when it comes to education and health care these anxieties melt away. We fondly imagine that a state which cannot bake bread can none the less run schools and colleges, hospitals and clinics. We are happy to entrust impressionable young minds and sick vulnerable bodies to state institutions.

When anyone dares to question these cherished rights, we appeal to social justice. Surely, we cry, the children of the poor should enjoy the same education as the children of the rich. And it would be obscene if someone were denied proper medical treatment because their bank balance was low. In such apparently unanswerable arguments, we forget that when it comes to food, shelter and warmth, which are even more crucial to health and well-being, the poor enjoy no such protection. And, more poignantly, we ignore the enormous disparity of standards between schools in poor areas and those in rich districts.

Our attachment to public education and health care goes far deeper than concern for justice; and, far from being inconsistent with the ideology of the free market, it is a necessary adjunct to it. At the time of the Industrial Revolution it was education, rather than health care, that attracted the greatest attention, and Adam Smith was joined by almost every other major economist of the period in arguing strongly for state involvement. Their main contention was that education is essential for economic growth. They recognised that the peasant in the village possessed a wide range of skills, from making a hay-stack to building a wall, which he acquired quite naturally through working with his parents. But once people arrived in the industrial cities these informal methods of education collapsed; the children were left alone to play aimlessly in their hovels while both mother and father went out to work. As Adam Smith grimly foresaw, a nation of intelligent craftsmen would be transformed within one or two generations into a race of ignorant morons. The dreams of limitless prosperity promised by industrialisation would thus be turned to dust, as the young men and women raised in the cities were unable to operate the factory machines.

Some more ardent proponents of the free market asserted that market forces would provide schools for the working classes. But Adam Smith and his followers were convinced that such hopes were misplaced. Education, they argued, is an investment in which the child may be regarded as 'human capital'; and the returns on the investment are not enjoyed for one or two decades—until the child reaches maturity. The working classes could not afford to make such an investment, since their entire income was needed for immediate consumption. And even if they could afford it, they would not necessarily make it, since most of the benefits of education accrue to the economy as a whole rather than to the individual. In Smith's view the extra wages that a worker would earn as a result of education were negligible; but if the work-force as a whole were educated, and thus able to pick up new ideas and techniques quickly and easily, the rate of growth would accelerate, to the benefit of all.

By the middle decades of the nineteenth century most governments in Europe were subsidising education. And at the end of the century the extraordinary upsurge of the German economy was popularly explained by Bismarck's heavy investment in technical colleges. In our own century vast state spending on schools and colleges has come to be taken for granted. And the economic arguments of Adam Smith are tirelessly repeated to justify the erection of yet another huge polytechnic or university, or to support the replacement of small village and neighbourhood schools with grand comprehensives, in which a wider range of subjects can more efficiently be taught, with the most modern technology at the children's elbows.

The development of state health services has been far slower, since the economic arguments are less obvious; but in the end the economic and political logic has proved equally forceful. If education is an investment, in which the child is likened to a machine, then health care is the repair of the machine. A healthy work-force, it is often said, is a productive work-force. And just as people cannot afford to make the initial investment, so the repairs too may be beyond their means, especially as the size and timing of the repair bill is unpredictable. Hence the state must pay for the repairs. The analogy with education goes even further. Just as the benefits of education accrue to society as a whole, rather than to the individual, so do the benefits of health care, especially in relation to contagious diseases: all of us benefit if the spread of such diseases can be checked by prompt medical treatment. So, since society as a whole reaps the reward, society through the state must pay.

Yet beneath all the economic arguments for public education and health care lie a much deeper moral attitude, which explains not only the fact of state provision, but also the form it takes. We have been happy to accept the image of a person as a piece of capital—with education as the initial investment, and health care as the subsequent repairs—because this image accords with our own convictions. By dividing the body from the soul, by splitting the public from the private aspects of our life, we have accepted the view that the body is no more than a collection of organs, a highly complex machine, whose sole purpose is to operate as efficiently and reliably as possible. The human intellect is the control box—the computer—which runs the machine. We recognise, of course, that the human animal has emotions and passions which are not mechanical and that the brain is capable of imagination as well as reason. But the world of passion and imagination has a quite separate orbit, centred on family life and friendships, and, for the minority, on the arts also.

Schools and hospitals are perceived as belonging, to a great degree, to the public sphere. When we visit a clinic, or become a patient in a hospital, we adopt an attitude of passive obedience, handing our bodies over to the doctor as we hand a faulty kettle to an electrician for repair. As Jonathan Miller has observed, the ritual of changing into pyjamas as soon as we enter the hospital reinforces this attitude—as does the very word 'patient' to describe our status. In our role as tax payers and voters, therefore, we judge our clinics and hospitals by their efficiency as workshops repairing the human body. And if the government assures us that closing down small local hospitals and opening large central hospitals instead will enhance efficiency, we do not quibble.

When it comes to schools and colleges, many teachers and lecturers rebel at the notion of themselves as all-knowing experts and their pupils as passive recipients of knowledge. They rightly perceive that education is most effective when the pupils are actively engaged in teaching themselves. And much effort has been expended in recent years to develop classroom techniques to encourage this active engagement. But for the two centuries since the Industrial Revolution we have not questioned whether a school, in which large numbers of children are artificially herded together, is the right environment for education. In a world where production occurs in large factories, we take for granted that education should be provided in large institutions. And if the parents must spend their entire day in a single place doing a single type of work, children are expected to do the same. Indeed, as Adam Smith recognised at the outset, if work is divorced from home, education must be also, since the alternative is to leave the children untended; and this in turn means that education must largely consist of classrooms of children passively following their teachers' instructions. The teachers who bravely promote active pupil involvement are fighting against their environment; and the lecturers at university, as their students unthinkingly note down their every word, know that the teachers have failed.

Those enlightened teachers who promote active learning, and the growing number of doctors who see the virtues of similar active involvement by patients in the treatment of illness, are in effect unearthing a profound contradiction in our approach to the young and the sick. We have by necessity placed education and health care in the public sphere, applying to them, as it were, the values of the body rather than of the soul. But we remain deeply uneasy. We recognise that historically the tasks of rearing the young and tending the sick belong to the family, and we feel that within the modern school and hospital our private values—the values of the soul—should temper the quest for efficiency. We wax sentimental about nurses, wanting them to be loving as well as productive; and we desire our teachers to look on their charges with the eyes of parents . Yet we know that love and dedication cannot be purchased for money—which is perhaps why in almost every western country we are content to pay nurses and teachers such low wages. And we are dimly aware that the way in which we organise public education and health care discourage such love.

PART TWO

SYMPTOMS

3

Ecological Pain

Earth's Rebellion

When Adam and Eve were banished from the Garden of
Eden, God condemned them to scratch a living from the
earth by the sweat of their brows: 'Through toil shall you
eat all the days of your life; thorns and thistles shall the
ground bring forth to you.' Outside Eden they no longer
had dominion over the plants and animals, but had to find
their place amongst them.

And that was how our species existed until about 7000
years ago. Only about three million of us were scattered
across the face of the earth, and we survived by hunting
wild animals, gathering wild roots and berries, and
sowing and harvesting the various wild grass seeds that
were edible. We were the most intelligent of the animal
species, and our front paws had evolved into hands of
exceptional agility. But we were by no means supreme,
and we were more vulnerable to attack than many larger,
stronger species. In short, we fitted neatly and unobtru-
sively into the natural order.

Then around 5000 BC a tribe of humans living at the
head of the Persian Gulf discovered that by breaking the

49

banks of the rivers and digging ditches to irrigate the fields, their harvests of grain could multiply. The population in the area expanded rapidly, and soon officials were appointed so that the ditches were maintained and the water was allocated fairly. Small towns sprang up where grain and other foodstuffs, as well as tools and utensils, were bought and sold. Gradually the ditches were extended and trees were felled to bring more land under cultivation and to provide timber for the growing number of houses. And thus, to the best of our knowledge, the first civilisation was born.

But in creating a civilisation, our ancestors brought upon themselves the first ecological disaster. As the years passed the salts from the rivers poisoned the soil, destroying its fertility. And the felling of trees both reduced the rainfall and loosened the earth, so that the rich soil was turned into barren dust blowing in the wind. Within a few centuries the people had fled to seek new pastures elsewhere, leaving behind them a desert—which it remains to this day.

Since that first civilisation, the same grim story has been repeated innumerable times across the world. The bright white sunlight of Greece, gleaming on the rocky landscape, is a reminder of an ecological disaster wrought by the wise democrats of Athens. They denuded the hillsides around the city to provide timber for houses and temples, leaving the torrential winter rain to wash the soil away. Then the trees around the rest of the country were felled to make way for olive groves; but the olive tree, with its deep taproot and dearth of horizontal roots, also left the soil unprotected. At the same time in Afghanistan one of the mightiest empires the world has known—now largely forgotten—was creating an irrigation system of extraordinary sophistication: thousands of square miles of lush forests were cleared to make way for fields of wheat and rice. But as with the original civilisation, the soil was soon rendered infertile, and the empire sank into the dust it had created. The barren wastes of Ethiopia, of western China, and of northern India, are also the remnants of ancient civilisations.

The greatest man-made desert, which continues to spread at an alarming rate, is the Sahara. Two millennia ago it was only a few hundred miles across, deep in the interior of Africa; and the plains to the north were the granary of the Roman Empire. Year after year, decade after decade, ships piled high with grain left the ports of North Africa to feed the towns and cities of the Mediterranean. But by the time the Empire fell in the fifth century, intensive farming, aided by grand irrigation systems, had destroyed the soil, bringing the desert right up to the sea shore. That was how it remained until the last century, when the spiritual heirs of imperial Rome sent traders and armies to the peoples on the southern edge of the Sahara. Tribes whose traditional methods of farming were ecologically exemplary, were brought under Western rule, and their territory ploughed up to provide food for the industrial cities of Europe. Now, despite political independence, the nations of sub-Saharan Africa have little economic choice but to continue the Western pattern, selling crops for cash on the international market. As a result the Sahara creeps southwards by one or two miles each year, and almost permanent famine stalks the land.

So the pride of the human animal, which seduces him into imagining that he can dominate Nature, is as old as civilisation itself. And the deserts of the world bear silent witness to our fearful folly. But the aspirations of our ancestors are small compared with those of our own era: the scientific revolution has carried human pride to ever loftier peaks of folly. Not content with bringing the soil under our control, we drill deep beneath it to extract all the riches of the earth's crust. Not content with spades and axes, we now arm ourselves with machines of unimaginable power and intelligence in our battle against Nature.

But of course, the higher our pride lifts us, the further must be the eventual fall. In the late 1960s, when the ecological prophets first sounded their trumpets, the problems seemed bad enough. They warned us of the imminent depletion of fossil fuels and other mineral resources; they alerted us to diseases caused by industrial toxins; and they lamented the filthy rivers and lakes in and around our great cities. We were severely jolted, but we perceived no fundamental threat to our way of life. We dared to hope that a few well-organised campaigns would prompt Western governments into developing alternative sources of energy and imposing anti-pollution regulations on offending industries.

Now, however, even schoolchildren can recite a mind-boggling list of threats to our very survival. The oppression of Nature by human science and technology is proving so acute that Nature seems to be rising up in revolt, turning its most powerful forces against us. As our aerosols and refrigerators destroy the ozone layer in the upper atmosphere, so the sun's rays, once welcomed as benign, now inflict fatal cancers. As our power stations belch out sulphuric fumes, rain showers, once hymned as soft and refreshing, now contain acid that destroys lakes and forests, killing the fish and wildlife that inhabit them. And as our roads seethe with cars and lorries, their exhaust turns the atmosphere itself into a thermal prison, trapping the sun's heat. The children who today draw colourful diagrams of the greenhouse effect and acid rain in their exercise books will witness in the course of their lives the greatest transformation of the environment in the entire history of our species.

The extent and the effect of that transformation is now a matter of earnest debate, with scientists offering widely differing views. But all the scientific speculation points to a single conclusion: that if our civilisation remains unaltered, the human species will in the course of the next millennium become extinct. We are already destroying many thousands of the other species with whom we share this planet, in each case bringing to an abrupt conclusion many millions of years of evolution. Our own demise is now imminent.

If we do disappear, our extinction will, in quite chilling fashion, recall the extinction of an earlier creature which once strode supreme upon the earth, the dinosaur. It has long been a mystery why this great reptile suddenly disappeared, but recently a remarkable and quite plausible explanation has been suggested. According to this theory, a great meteorite, perhaps five miles across, hit the earth. As it entered the earth's atmosphere it became red hot, setting alight the oxygen and nitrogen above the earth. For a few moments the sky became a sheet of flame, turning the nitrogen into nitric acid. The meteorite landed in the sea, sending a huge tidal wave across the world, flooding low-lying land and salinating the soil. As it hit the water the meteorite broke into tiny pieces, throwing up a huge cloud of dust. For three months this cloud covered the earth, plunging it into icy darkness. The reptiles which survived the flood now died of cold. The trees were freeze-dried, and in a thunderstorm, when lightning struck a tree, a great fire swept across the continents, leaving a layer of soot which can still be found. And the rain brought down the nitric acid, poisoning rivers and lakes, and so killing fish and birds. In the meantime the dust-cloud was destroying the ozone layer; and the great fire hugely increased the level of carbon dioxide in the atmosphere, creating a greenhouse effect which raised the earth's temperature and sea level.

The planet, of course, recovered: some species survived, new species such as *Homo sapiens* evolved, and ecological balance was restored. But it took many millions of years.

Salination of the soil from irrigation schemes, acid rain from power station pollution, a temporary ice age caused by the dust of a nuclear explosion, the depletion of the ozone layer from CFCs, a greenhouse effect from carbon dioxide emissions—the threats to our own survival are a ghostly echo of the ecological cataclysm that might have destroyed the dinosaur. There is, of course, one fundamental difference. The dinosaur was a helpless victim, which neither caused the catastrophe that befell it nor could have done anything to prevent it. Humankind, by contrast, is the author of the ecological disaster that now looms. The question which people in every part of the world are now at last asking, with fearful urgency, is: can we, even at this late stage, prevent it?

Illusion of Wealth

When we try to analyse our ecological mess, we quite naturally turn to science. It is, we feel, science which first caused the mess, and so science should be able to explain it and offer solutions. Thus we are all now familiar with long articles by eco-minded physicists, chemists and biologists calculating the billions of tons of sulphuric acid, carbon dioxide and other pollutants which we pump into the atmosphere each year, and predicting with the aid of elegant diagrams their effects on the natural environment. It is remarkable how quickly we have all become armchair ecologists.

But to confine our attention to science is to miss half the story. Scientists do not work in a vacuum, but within a social, political and economic system which both provides them with resources and harnesses their discoveries. Science itself can justly claim to be neutral; it is the system in which it operates that determines whether its effects are benign or malign. The tragedy of our civilisation is that both the dominant ideologies, capitalism and socialism, encourage the malign potential of science.

The free market, by its nature, can only take into account the needs and desires of the individual consumer; it cannot meet collective needs and desires. Thus it is extremely efficient at supplying food, electrical goods, clothes, and the like: a potato, a fridge and a shirt can only be used by one person or family. But it cannot put up street lights, provide an army or police force, build roads and bridges: these benefit society as a whole, and so the individual acting alone cannot go out and purchase them. This deficiency of the free market was recognised by Adam Smith, and he believed that the state should supply these public goods, paying for them by raising taxes.

Smith, however, did not recognise a similar deficiency in relation to producers. The individual firm, when deciding how to organise its factories, will take into account all costs for which money has to be paid: the costs of labour, raw materials, rent for the factory building and so on. But it will not take into account costs which it imposes on the community at large. The most important and obvious of these social costs is pollution. Within the free market the firm does not pay to pump its waste into the nearby river, or to push toxic gases up its chimneys. A more hidden social cost is the depletion of natural resources: free-market non-renewable resources are priced according to the financial cost of their extraction, but this ignores the cost to future generations of not having those resources.

Apologists for the free-market often put their faith in the goodwill of the producers, believing that their sense of responsibility to—and perhaps desire for respect within—the local community would prompt them to curb the pollution. And today companies spend many millions on advertising to assure the public of their green credentials. People of strong green convictions also sometimes imagine that public pressure on companies can induce them to mend their ways. To a limited degree such pressure can work, especially if members of the public in their role as consumers can be persuaded to forswear ecologically unfriendly products.

But such pressure misses the fundamental point: that even if the directors and managers of all our companies were ecologically enlightened, acting as individuals they are powerless. The free market compels individual firms to produce their goods at the lowest possible financial cost or else be driven out of business by their rivals; and if this requires high levels of pollution, then the individual managers must turn a blind eye.

The dilemma is best illustrated in our most basic industry, agriculture. Many farmers genuinely wish to conserve the countryside. But they operate in a highly competitive market with very narrow profit margins. A small number can farm organically, selling their produce at a higher price to the minority of consumers who are prepared to pay the premium. But while the majority of consumers simply want to buy their food as cheaply as possible, the majority of farmers must keep their financial costs to the minimum. If this requires large doses of chemicals and heavy machinery, the individual farmer must buy the chemicals and the machines—or else run at a loss.

The same applies to our entire economy, from steel production to the manufacture of computers. Of course there are some firms which cause unnecessary pollution which could be curbed without financial cost. And some environmentally harmful practices achieve only short-term profits, while jeopardising long-term prosperity. The most notorious example is the felling of the rain forest in the Amazon basin, where the new cattle ranches survive only a few years before the soil is exhausted. But, as any financial broker will confirm, the market-place has a frighteningly short time-horizon, so the lure of large immediate profits overwhelms all anxieties for the future. The cattle ranchers of Brazil and the farmers of Europe are controlled by forces far more powerful than any moral scruples they may possess. And those same forces push toxic gases out of factory chimneys and rape the earth of her precious minerals.

Socialists might reply that the socialist state embodies this collective interest, controlling its economy on behalf of society as a whole. Thus in principle the state should be able to prevent pollution, operating its factories cleanly. Unfortunately the evidence of socialism in practice gives little cause for optimism. Socialist control in Eastern Europe from the late 1940s to 1989 created the vilest economies the world has yet seen. Sewage is poured untreated into lakes and rivers; ill-conceived factories emit all manner of poisons; unsafe nuclear power plants threaten not only the workers within them but their children also; and their cars are so poorly designed that their exhaust emissions outstrip those of their western neighbours, despite the tremendous disparity in car ownership. Parts of Eastern Europe are an ecological nightmare in which the damage the human race is doing to this fragile planet is made horrifyingly clear. For the first time in many centuries life expectancy has fallen dramatically, as children and adults alike are struck down by the new range of diseases caused by pollution.

It is all a far cry from the high ideals of the socialist pioneers, who fervently believed that a state which owned the means of production could eradicate all collective ills. There are many socialists today who assert that the disasters of Eastern Europe are an unfortunate aberration, which can be corrected by a few adjustments to the socialist machine; but the stark truth is that socialism, like capitalism, is by its nature indifferent to the environment. And the reason for the indifference is identical: those who manage our resources are accountable to no one.

In the free market it is the individual firm which enjoys absolute control of the use of resources. If a farmer owns a stretch of land, it can be cultivated as he or she pleases. If a mining company suspects that there is coal beneath a range of hills, it can purchase the necessary rights and start digging. If a domestic appliance company decides that refrigerators can be made most cheaply and

THE HEALTH OF NATIONS

efficiently using CFCs, then it will order those chemicals. If the internal combustion engine is the most convenient and fastest source of locomotive power, then our 'horseless carriages' shall all burn petrol.

Under socialism there is a great bureaucracy which manages resources. Central planners set production targets for farms and factories, and they determine how those targets shall be met, allocating raw materials and machinery accordingly. The managers of the various enterprises merely put those plans into practice. The jobs and the salaries of the planners and the managers do not depend on the efficiency of the state enterprises, so there is a notorious misuse of resources under socialism. Moreover rigid price control makes it easy to disguise inefficiencies, so they can escape any public pressure for improvement. Even if, as socialist faith requires, planners and managers were dedicated to the public good, they would still have an impossible job. They would have to make millions of consumption and production decisions which in a free market are made by individuals acting for themselves.

The vile pollution caused by socialist economies is not due to ill will; as under capitalism it is due to indifference, which is intrinsic to the system itself. If a vast chemical factory is spewing out a poisonous gas over the local population, there is no individual, or even a definable group of individuals, who can be held responsible; and there is no individual or group with any incentive to put things right. Outside observers visiting socialist countries are often surprised by the apathy of the people, who suffer in dumb silence the diseases and discomforts caused by pollution. Yet the people have long since learnt that protest is at best ineffective, because there is no one to whom to direct one's complaints, and at worst dangerous, since socialist states frequently stifle dissident voices.

The modern apologists for socialism often argue that it is not socialism as such but lack of democracy which is at fault; and they bitterly criticise the post-war socialist regimes which, having gained power, lacked the courage to hold popular elections. If, they contend, the central planners—or rather their political masters—had to submit to the ballot-box every few years, then the desire to retain office would spur them to greater efficiency and environmental controls. Yet democracy, even at the local level, is an extremely crude and blunt instrument for ensuring accountability. The issues involved in managing an economy and protecting the environment are often extremely complex and defy the simplistic arguments on which elections turn. More fundamentally, parties seeking election submit an entire package of policies to the voters; thus the voters cannot express their views on each particular policy. Certainly socialism based on democracy is likely to be far more benign than authoritarian socialism; but it is also likely to show exactly the same inefficiency and ecological indifference.

Yet, although capitalism and socialism share a basic flaw, there are differences which are important when we come to consider an alternative. Karl Marx himself acknowledged that capitalism was the most dynamic economic system which the human race had ever invented, achieving rates of growth and levels of efficiency beyond the wildest dreams of our feudal ancestors. He saw that its secret lay in harnessing the natural self-interest of individuals, which had hitherto been largely wasted in warfare and political rivalry, towards the pursuit of wealth. In this respect he was entirely correct; and any system whose aims include the conservation of the natural environment must possess the same dynamism. At the same time, a new system must imitate socialism in having a strong central authority which can express and enforce collective aims and policies. Pollution can only be controlled, and resources can only be conserved, if we learn to act in unison. The great challenge, therefore, is to devise a political system in which the self-interest of individuals can be channelled towards the collective conservation of the environment.

4

Social Pain

Soul's Rebellion

Karl Marx observed that in the industrial city a man only feels at peace with himself when he is at home. When he is sharing a meal with his family, digging his garden, playing with his children or repairing the house, the two aspects of his life, the spiritual and the material, are unified. At home he lives and works within a close spiritual bond, that of the family. As he performs his various household duties he is sustaining and strengthening that spiritual bond; and the bond itself gives meaning and purpose to the most menial and mundane tasks. But when he leaves home in the morning to spend the day in a factory or office, his work has no spiritual dimension; on the contrary, he finds himself in a large crowd whose only common aim is to make money. He thus feels restless and uneasy, his soul empty and dissatisfied.

Marx coined the term 'alienation' to describe the spiritual plight of modern humanity. The individual is alienated in two respects. First we are alienated from our fellows, since on the factory floor or in the office we are isolated. We may strike up pleasant conversations with other workers, but such friendships remain superficial because they lack any comon material purpose. Secondly we are alienated from work itself, since the task each person performs is only one tiny part of the whole process of production; the link between individual effort and final product is therefore invisible. Moreover, we are likely never to meet the consumer who is the beneficiary of our work.

The workers, according to Marx, for the most part tacitly accept their condition, since they fail to connect their inner sense of emptiness with the outward cause within the capitalist economic system. And capitalism itself provides the means of dulling their spiritual discontent. Under capitalism, workers are, in Marx's view, 'commodity fetishists', striving to purchase increasing quantities of material goods to compensate for their alienation. Goods become a drug, whose dosage must rise to maintain the illusion of well-being. At some point, he predicted, the fantasy will shatter, and the horrifying truth will dawn—and then the revolution will begin.

Fifty years after Marx, in the opening decade of this century, Emile Durkheim conducted a similar psychological study of industrial capitalism, in which he focused on the moment when the illusion shatters. Durkheim contrasted modern industrial society with that of the primitive tribe. In a tribe the individual lives within a close-knit group, which instils its own values and norms, governing every aspect of daily life. In some tribal cultures the individual is so fully subsumed within the group that the tribe lacks any pronouns in the first person singular: they speak not of 'I', 'me' and 'mine', but always of 'we', 'us' and 'ours'. The spiritual and material realms are thus completely integrated.

In industrial society the individual stands virtually alone, compelled to decide for himself what goods to consume, what work to do and where to live. Such corporate values and norms as exist are weak and flimsy, easily overridden. The dominant value thus becomes the prosperity of the individual. For many the pursuit of material wealth proves a sufficient motivation, enabling them to stay at their factory bench or at their office desk. But for some the drug loses its potency, and they plunge into despair. Durkheim called the despair 'anomie'. It is quite different from the occasional bout of depression or frustration which all human beings experience, even within a tribe. The despair of anomie reaches down to the very heart of our existence, since the individual has to confront the spiritual void at the core of society itself.

To Durkheim the ultimate response to anomie is suicide, since life without social and moral values is untenable. In a tribal society suicide is unknown, whereas in Western industrial society it is relatively common; and Durkheim regarded the suicide rate as a measure of the degree of moral breakdown. But he saw other less extreme symptoms of anomie. The most obvious are proxy forms of suicide, such as addiction to alcohol and other drugs, where the individual is trying to destroy his or her own soul. Mindless and violent crime, such as vandalism, is also a sign of anomie, and so also are certain forms of mental disorder. As Durkheim emphasised, these are all tips of a very large iceberg, since the despair within Western society is largely hidden and unexpressed.

Though Marx and Durkheim were writing a century or more ago, we can all recognise our own social conditions in their descriptions. Our homes are more than ever the sole focus of intimate relationships; and beyond the home, bonds are weak or non-existent. Many of us continue to feel alienated from the work we do, finding little personal fulfilment in our jobs. Our addiction to an ever rising degree of affluence is stronger than ever, as last year's luxury becomes this year's necessity. When ecologically- minded economists speculate about an economy without growth, the idea is quickly discarded as a psychological impossibility: we are as hooked on growth as the addict is on heroin. And beneath it all there is a strong undercurrent of despair. Suicides among teenagers and young people have almost doubled in the past fifteen years. Alcoholism and other forms of drug addiction continue to increase, despite huge campaigns alerting us to the dire effects; indeed, if these are forms of suicide, then such campaigns may make them more attractive. Violent crime also rises inexorably, despite harsher punishments.

If Durkheim were writing today, he would probably regard mental disorder as a better index of our social breakdown than suicide. Our most widespread mental problem is stress, which in turn causes a wide variety of physical diseases, some fatal. Stress is to a great extent a symptom of alienation. In our desperate attempt to deny the inner emptiness of our lives, many of us work ridiculously long hours at a furious tempo, chasing ever higher salaries: the addiction to adrenalin combined with material luxury can mask for many years our inner despair. But when our careers judder to a halt, or when mind and body break down under the pressure, then the mask slips. Stress is one measure of our alienation; so the eventual nervous breakdown, which often leaves the victim chronically depressed, is a further index of anomie.

Karl Marx in a famous phrase referred to religion as 'the soul of a soulless society ... the opium of the people'. This is usually taken as denigrating religion, but in fact Marx's meaning was more subtle. Faced with alienation, people search for things to make life bearable. Material luxuries are the most common refuge, but religion is more powerful because it directly addresses our spiritual dilemma. According to Marx, religious institutions provide a substitute community, a kind of pseudo-tribe to compensate for the lack of real social bonds. The most potent religious 'tribes' are the innumerable sects, where for two or three hours each week working people can sing heart-warming hymns and choruses, rejoicing in their spiritual fellowship on earth and looking forward to the perfect community of heaven. Marx did not despise this form of religion, because he understood people's psychological need for it. But he believed it was merely blinding people to their misery on earth, offering them hope beyond the grave when they should be fighting for justice on earth.

Durkheim—who like Marx was an atheist—had an even more positive view of religion: he saw it as the cement of society. Religious symbols and rituals are the means whereby society's moral and social values are conveyed to its members, giving meaning and purpose to their lives. Thus the dances, the festivals, the totems, the sacrifices and all the other paraphernalia of a tribal religion hold the people together; and by zealously preserving their religious traditions, the tribe passes its values from one generation to another. The decline of religion in Western society is both a cause and a consequence of alienation and anomie. The Protestant Reformation deprived Europe of its religious cohesion, both by breaking the ancient Church in two and by turning religion inwards, focusing on personal salvation rather than social well-being. This encouraged the growth of capitalist economics, which exalts the individual; and capitalism in turn left no place for religion, because by its nature it undermines common values. Durkheim believed that western industrial society could only be lifted from its misery by the re-emergence of religion as a powerful moral and social force. But his own intellectual predicament convinced him of the futility of such a hope: as an atheist he regarded all religion as objectively false, and so could not conceive how people today could collectively be induced to accept a fantasy. In short, he wanted the outward symbolic trappings of religion without the inner faith—and he knew that was impossible.

Today religion continues to stand on the edge of society. While most people, according to opinion polls, continue to affirm their belief in God, only a small minority attend worship regularly. Although the founder of the Church preached a revolutionary social and moral vision, his disciples today fit easily into society, threatening no one and upholding established order. Religion, in the eyes of the majority, is a harmless hobby. In one respect Marx's view no longer applies, and to a great extent did not even apply in his own day: if religion is the opium of the people, one would expect to see the poorer classes attending the churches and chapels, while the rich stayed comfortably at home. In fact even in Marx's time it was predominantly the middle and upper classes which were devout, while the working classes felt as alienated from institutional religion as they did from their jobs. This largely remains the case today. The explanation, however, is not hard to find. The majority of people—and most especially the lower classes—compensate for their alienation by the pursuit of affluence, while those already satiated must look elsewhere—and religion is an attractive opium.

In recent years a new, secular religion has sprung up, which seems to offer the rewards of conventional religion without its supernatural beliefs. This secular faith has many different forms—sects, as it were—but can broadly be labelled psychotherapy. Like Protestant Christianity its focus is the feelings and emotions of the individual, and its purpose—like Protestantism—is to replace unpleasant, negative feelings with warm, positive ones. In its myths and methods it draws heavily on Freud and Jung, seeking the source of negative emotions in childhood experiences, and encouraging the individual to remember and relive those experiences in the hope that this will cleanse the psyche. Yet to anyone familiar with Christianity—or indeed any religion—the similarities are striking. While professing to be free of moral values, the labelling of emotions as negative and positive sets up an alternative theology of sin and righteousness. Almost all forms of psychotherapy regard guilt as a major burden on the psyche, and seek to assuage it. In Christian terms they are offering 'redemption'. The psychotherapists consulting-room is the confessional by another name, while group psychotherapy is familiar to monks and nuns as the 'chapter of faults', with the psychotherapist acting as abbot or abbess.

Religion, both in its conventional and its psychotherapeutic forms, gives great help and comfort to individuals. If it is an antidote to much human despair, and if it sustains many people in their daily round, it should not be despised. And, looked at purely as a compensation for human alienation—Marx's 'soul of the soulless society'—it does no damage, while the alternative, 'commodity fetishism', continues to lock us on to the treadmill of economic growth. Most important of all, the religious tradition contains within it the seeds of a new vision: seeds which have remained largely dormant, but which are being carefully preserved.

The seeds will only germinate and burst into bloom, however, when the Church can escape the dark shadow of Platonic dualism. As Durkheim understood, the symbols and rituals, the myths and prophecies of religion are potentially as powerful today as they were in the past. But religion itself, by dividing the spiritual realm from the material, has hidden those symbols, rituals and myths from light. The churches invite people to step inside and receive spiritual salvation and emotional comfort; and the priests and ministers are more expert today than ever before in conducting worship and expounding theology. But even though many priests and ministers fervently wish that it were otherwise, their worship and theology do not spill over into the street outside and onward to the factory and office. While religion remains private, it is impotent.

Illusion of Care

Marriage is more popular than ever, yet the rate of marriage breakdown is higher than ever. At first sight those two bald facts about modern society appear contradictory: surely, one might ask, if marriages are so vulnerable, and if the high rate of breakdown signifies widespread misery within families, people should be deterred from marrying. But in truth they are two aspects of a single dilemma—which is itself symptomatic of the wider social and moral malaise.

Prior to industrialisation people expected little of marriage. It was essential both for procreation and for social cohesion, but it represented a bond between two families as well as between two individuals. Husband and wife sought companionship outside the home as much as within it, so even if they had little in common, they found ample emotional fulfilment within the wider network of relationships. Divorce was rare, not only because economic and moral pressure forced couples to remain together, but also because it was emotionally unnecessary. To a surprising degree sexual immorality was tolerated, at least among men, where a marriage proved unsatisfying. The great theologian Thomas Aquinas even regarded brothels as a necessary, if unfortunate, support for the marriage bond: better to resort to a prostitute than to divorce.

But in the nineteenth century, as the new industrial cities sprawled across the countryside, both sentiment and morality were transformed. As married couples found themselves isolated, cut off from the ancient ties of the extended family, so romantic expectations rose. The home was now the sole focus of emotional fulfilment. Popular songs extolled the blessings of matrimony, offering each new generation the dream of passionate romance and life-long love. Sexual ethics became strict and rigid: sexual activity before marriage, even of the mildest form, was forbidden, adultery was unmentionable and divorce was inconceivable. This marital revolution mirrored the Industrial Revolution; and as the emotional isolation of the family became more complete, so the romantic ideal was more fervently praised. Marriage was the luxuriant oasis in a spiritual desert.

Yet as the ideal was lifted higher, so reality sunk lower. Novelists of the period grimly record the misery and oppression of many nineteenth century homes. The isolated marriage could rarely bear the weight of expectation placed upon it, so many experienced matrimony not as a source of happiness but as an emotional prison. And the gap between morality and behaviour grew ever wider: by the latter half of the nineteenth century any man who could afford it was virtually expected to seek sexual solace outside the home. The rich kept mistresses, while the less affluent went to brothels—Thomas Aquinas's dictum was applied with a vengeance, as the oldest profession boomed as never before. Of course, there were numerous marriages which were happy and harmonious, where husband and wife accepted the limitations of matrimony. But for society as a whole marriage and family were entering a prolonged period of crisis.

Today the romantic ideal is more powerful than ever. The records bought by adolescent boys and girls continue to extol life-long monogamy as the only route to happiness. Small fortunes are now spent on lavish wedding ceremonies, in which bride and groom appear as lovers in a fairy-tale, destined to live happily ever after. Far from being cynical, the offspring of divorced parents are the most fervent in their marital hopes. And the divorcees themselves, far from giving up in disgust, continue to search for the perfect partner, leaping eagerly into second, third and fourth marriages—although in many cases the new partner is disconcertingly like those previously discarded.

Much of the old hypocrisy has, however, been stripped away. If hope continues to triumph over experience, at least we are honest when hopes are dashed. Indeed the high divorce rate is not so much a sign that marital misery is greater than a century ago, but rather a willingness to acknowledge that misery openly. Although hard facts in such areas are notoriously elusive, there is little doubt that prostitution has declined sharply, while adultery between freely consenting partners has risen. In the 1960s and 1970s people talked glibly about the 'permissive society' in which sexual morals had supposedly been abandoned. In fact sexual fidelity is as highly prized today as ever, not least amongst the young where 'sleeping around' is widely condemned. We simply talk more freely about our lapses.

Lurking in the gulf between shining ideals and tarnished reality lies our odd attitude towards homosexuality. In the nineteenth century homosexual practices were condemned more vehemently, and punished more severely, than in any previous period in our history. They were a manifest affront to the romantic ideal; and the anger with which homosexuality was attacked reflected the anxiety which that ideal aroused. Knowing their own hidden shortcomings, people lashed out at those who openly flouted their immorality. Today, with our greater honesty, we might be expected to tolerate homosexuality; and certainly our legislation has become more sympathetic. Yet, as homosexuals will affirm, the hatred and vitriol directed towards them is as acute as ever.

Beneath our homophobia lies a profound ambivalence. On the one hand, we cannot bear those who seem to mock the romantic ideal, because we ourselves are so insecure in our grasp of it; on the other, we secretly envy the freedom of the homosexual, who is unconstrained by conventional morality. The homosexuals are themselves in a moral limbo. Many would like to contract marriages, like heterosexuals, yet society forbids them from adopting heterosexual ethics, refusing to recognise and bless gay partnerships. Homosexuals thus find themselves encouraged to be promiscuous, compelled to live out the sexual fantasies of the heterosexual. The moral dilemma of the homosexual directly reflects the moral and emotional crisis within marriage.

But the crisis is not confined to the privacy of the home: it reaches outwards to the modern extension of the family, the welfare state. Just as marriage is more popular than ever, so public health care and education are more highly treasured. Yet just as marriage is in a prolonged crisis, so also is the welfare state. The reasons, however, are directly opposite. In our division between the private, spiritual realm and the public, material one, the welfare state has been put on the material side. We regard education primarily as an investment, equipping our children with the knowledge and skills to become economically successful. And we regard health care as bodily maintenance and repair. This materialistic attitude is now under growing attack. Moreover, by placing education and health care under the state, we subject them to the vagaries and complexities of economic policy;

and here too the welfare state finds itself attacked, starved of funds when it should be expanding.

The crisis of attitude is most manifest in health care. In the past two decades patients have voted with their feet: disillusioned with the mechanical treatment they receive at the large hospitals and clinics, they seek instead forms of therapy which embrace their whole condition, mental as well as physical. Many doctors have themselves confessed to profound unease at their approach, recognising that many of the diseases for which they merely prescribe drugs have a psychological dimension also. Indeed, there is alarming evidence that drug-based medicine, while curing particular ailments, actually worsens overall health by breaking down our natural capacity for self-healing. When we compare one country with another, we find that while such basic facilities as good sewage disposal and clean drinking water extends life expectancy, the abundance of medical drugs can actually reduce it. In short, the strict division of body and spirit on which so much modern health care is based is both false and dangerous.

Doctors who acknowledge this falsehood are now striving to improve state health care, adopting methods pioneered in the various alternative forms of treatment. While enjoying limited success, they find themselves fighting against the entire ethos and organisation of state health care. Working in large clinics and hospitals, they have neither the time nor the opportunity to get to know their patients as individuals, still less their circumstances at home and at work. So they have no choice but to focus on the symptoms of disease, alleviating the immediate sources of distress as quickly and efficiently as possible. Until recently doctors have excused this approach by claiming that patients themselves would feel angry if they did not emerge from a consultation with some pill or potion. But now, as many doctors admit, the patients themselves are rebelling.

In education the dilemma is less acute because the effect of school on a child's eventual character and well-being is less easy to detect. Moreover in a highly competitive society parents are naturally anxious that their offspring should be well armed for life's battles; and for want of any other tangible criteria, they are apt to judge schools by examination results. But every parent and teacher knows that children can only learn successfully if they are in a happy and secure environment. Anyone who looks at the schools and colleges of the Western world—especially at secondary and tertiary level—with unblinkered eyes, must acknowledge a profound emotional malaise. Pupils and students are often bored and apathetic, while many teachers have long despaired of inspiring more than a tiny minority. In some institutions education is reduced almost to a daily struggle to maintain order and prevent chaos.

Unlike doctors, many teachers are striving to improve matters. Schools have been in a state of almost continuous revolution for the past two or three decades , as new methods and equipment are introduced. And whenever complaints are made, teachers and parents alike delude themselves that more money spent on computers, laboratories and videos will stop the rot. An odder plea, made with equal passion, is for extra money to be spent on teachers' salaries to attract better people into the profession. While there is much justice to the claim, it carries the awkward implication that the present teachers are intellectually inadequate. In truth the teachers should bear none of the blame for the problems of education; it is we, the voters, who are at fault for imagining that large institutions, managed by the state, can nurture young souls. A small number of tribal societies put their young into similar institutions for a few brief weeks at puberty, as part of their initiation into manhood. But no other culture in history has ever herded hundreds, even thousands, of children into one place, day after day over fifteen years, as a preparation for adult life. We have become so accustomed to such a system that we rarely question it; yet to anyone unfamiliar with Western education it must appear emotionally barbaric.

If we find it hard to look critically at our methods of education and health care, we have a similar mental block when it comes to the funding of these services. Since we require education and health care to be free or subsidised, they are financed by taxation. In the early stages of state welfare, when education for the majority finished at puberty, and when medicine was comparatively simple, the burden on state revenues was quite light. Today, however, we rightly extend education until full adulthood, and even beyond for those who wish it. And, as people live longer and the range of treatments widens, so we each need much more medical attention in the course of our lifetimes . Equally, as society as a whole has grown more prosperous, and as our homes are now cluttered with material goods, we naturally wish to devote a higher proportion of our resources to education and health care, demanding higher standards.

If we paid for these services through the market-place, then they would now be expanding rapidly. Just as increased demand for foreign travel has, through the market mechanism, induced an exponential growth in the tourist industry, so the same process would be occurring in education and health care. Indeed, if the analogy with tourism is correct, we could expect the numbers of people employed as doctors, nurses, teachers and lecturers to rise by three or four times in as many decades. But since we pay, not as consumers, but as tax payers, there is constant pressure to cut expenditure. Opinion polls reveal that we paradoxically want both higher spending on welfare services and lower taxes; but when it comes to casting our votes, as all political parties are aware, tax reductions prove more tempting. And even if voters were in principle willing to accept higher tax rates to pay for better services, such political altruism would soon prove self-defeating. Economics have long recognised that raising taxes above a certain rate actually reduces total tax revenue, since the incentive to work— and thus earn money on which to pay taxes—is stifled. It would seem that in most Western societies taxes are already at, or even above, that rate.

The welfare state is thus caught in a trap from which there is no escape. It cannot expand, and is condemned to be constantly squeezed by politicians eager to cut taxation. At the same time demand is constantly growing. Not surprisingly there is rising public frustration in most Western countries at the shabby conditions of hospitals and schools, and at the poor service provided in them. Yet we imagine ourselves to be so dependent on the welfare state that we remain blind to the cause of our frustration. The tragedy is compounded when one recognises that health care and education are the two most ecologically friendly activities on which to devote our time and money: teaching a child or nursing an invalid harms nothing.

PART THREE

CURE

5

Natural Harmony

Spirit of Creation

The Book of Proverbs in the Old Testament speaks of how God created wisdom at the beginning of time, and how wisdom became the architect of the entire universe. God's wisdom can thus be discerned within every mountain and lake, every animal and bird, and every human being; through this wisdom God is present in everything.

At about the same time that Proverbs was being written in Israel, the Stoic philosophers of ancient Greece were formulating a similar outlook. They coined the term *logos*, meaning word, to describe the power of God in every object: the logos is the essential quality of an object, such as the hardness of a stone or the sheen on silver. They also spoke of *pneuma* as the motive and energy of an object. Applied to humans, the logos is our rational intelligence, and the *pneuma* is the will. The purpose of human life, according to the Stoics, is to achieve perfect wisdom. This means that a person's *logos* should be in complete accord with the *logos* throughout all creation; and our *pneuma*, and thus our practical actions, should be in perfect harmony with the activities of Nature as a whole. The truly wise person will never wantonly cause injury to other creatures, including animals and plants, since that would upset the natural harmony of creation. Instead every action will be directed towards the well-being of all.

The evangelist John brought together the Hebrew and Greek philosophies in the opening verses of his Gospel. He adopted the Stoic notion of *logos* to describe God's agent in creation: *logos* 'was in the beginning with God: all things were made through him, and ... in him was life'. Far from having withdrawn from his creation, God remains present and active, guiding and inspiring his creatures. And the central principle of *logos* is love, revealed fully in the person of Jesus Christ. Thus the wise person will not seek to dominate his fellow creatures, but to love them, striving for perfect harmony between all living beings.

Sadly the theologies of the early Church quickly lost sight of John's universal vision, concentrating instead on the narrow religious issues which his theology raised. They were anxious to prove that Christianity was superior to all other faiths, and thus debated at inordinate length how the person Jesus Christ could embody the divine *logos*. Platonic dualism soon reared its head, with many arguing that a human being, made of flesh and blood, could not be unified with the divine spirit, since body and spirit are in perpetual warfare. Some even argued that the historical figure of Jesus in Palestine was a mere phantom, a kind of ectoplasmic illusion through which *logos* became visible. Finally an awkward compromise was reached in which Jesus was perceived as being two spirits, divine and human, which co-existed in a single body. By implication, all other men and women, possess only a human spirit; and animals have no spirit at all. Unfortunately this missed the entire point of the Stoic and Hebrew philosophies which lay behind John's notion of *logos*, that the divine spirit exists within all creatures. Not surprisingly, when the early theologians turned their attention to animals they saw them merely as existing for man's convenience, to provide meat, butter and fur. This has remained the dominant Christian attitude ever since.

There have, however, been some stout and notable exponents of *logos* theology over the centuries; and it is from them that we in the Western world should now seek a philosophy for our time. As early as the second century, Irenaeus, the redoubtable Bishop of Lyons, asserted that Platonism contradicts the Bible, which teaches that spirit and matter are indissolubly linked. He pointed to the vision of paradise in the Old Testament in which all creatures live together in harmony: 'The wolf shall dwell with the lamb, and the leopard lie down with the kid.' And in this vision the human is simply one creature among many, each glorifying God according to its own nature. All this, he observed, is a far cry from the Platonic vision, in which the soul escapes from the natural order. More poignantly, according to Irenaeus Platonism is incompatible with the resurrection of Christ, which is the heart of the Christian faith. At Easter it was not a disembodied spirit which rose from death, but body and soul together. To Irenaeus the Platonic idea of heaven as liberated souls existing in some immaterial space was mere fantasy. The heaven as promised by Christ's resurrection is of all bodily creatures living in perfect unity; and God calls us to establish this heaven on earth.

Irenaeus was a fighter, defending the truth as he saw it against attack. A millennium later his theology briefly caught the European imagination when it was taught and practised by a charismatic saint, Francis of Assisi. Francis was a wandering troubadour who sang the praises of God's creation, and urged everyone to bask in its beauty. His famous hymn to 'brother sun, sister moon' and his homily to the birds enjoyed huge popularity amongst the ordinary peasants. 'My little sisters,' he addressed the birds, 'your duty like ours is to bless God everywhere and always ... for the good he provides you, for the songs he has taught you, and for the air in which you fly.' Bonaventure, Francis' disciple, gave theological form to his master's poetry. Nature, he taught, precisely reflects the divine trinity: the power of natural life emanates from the power of God the Father; the wisdom which all Nature possesses, each creature living according to the divine laws, is the wisdom of *logos* as made flesh in his Son, Jesus Christ; and the goodness of Nature, providing ample food and comfort for all creatures, is the work of the Spirit, constantly present in every aspect of creation.

A century later the great German mystic, Meister Eckhart, made the concept of *logos* the cornerstone of his teachings on prayer and morality. Eckhart pictured human life as four stages on a journey. The first is to recognise that everything—every tiny insect and every small stone—is 'an expression of God's *logos*'; and so in looking at insects and stones, plants and animals, we are seeing embodiments of God himself. The second stage is to put this insight into practice, by caring for all living things, and treating the earth itself with reverence: just as a horse, when the tether is taken off, 'pours its whole strength into leaping about the meadow', so human beings, when 'the tether of ignorance is released, will want to love God's creation with all their energy'. The third stage is to 'let go' all desire to possess land or objects, because such desire will always make us want to exploit God's creation for our own ends; instead we should regard all things as belonging to God, loving them as God has made them. The final stage is for the individual to see himself as a prophet, called by God to proclaim these spiritual truths in the world of politics and business: 'What we plant in the soul through prayer, we must reap in the harvest of action.' Small wonder that in 1327 Eckhart was convicted of heresy—a fate from which Francis only narrowly escaped.

With the rise of scientific inquiry in the seventeenth century, and with the Industrial Revolution a century later, the stakes rose dramatically: no longer was it merely Christianity that was at issue, but the nature of civilisation itself. The majority of philosophers took the side of the scientists, regarding Nature as a giant machine, devoid of spirit, to be harnessed for human comfort. The most influential of these philosophers, Descartes, adopted Plato's dualism, picturing the human mind as standing over and above the material world, manipulating it as puppeteers control their models. A single philosophical voice, however, spoke out against the rest—that of Gottfried Leibniz. Leibniz refuted the mechanical view of Nature, and rejected the notion of the human mind standing outside Nature. He insisted that 'mind' is to be found in every particle of Nature, giving life and meaning to every object and every event. He portrayed the universe as a harmonious system in which there is unity amidst multiplicity, co-ordination amidst differentiation; and humankind is simply one part of this whole, one species amongst many. He warned scientists against seeking to dominate and manipulate Nature, urging them to respect the delicate balance within the complex natural order. If we in our arrogance upset that balance, we destroy that God-given harmony on which all species, including humankind, depend. Thus, in trying to control Nature, we will destroy ourselves.

In his day Leibniz, though respected by many, was regarded as an eccentric. His ideas, however, were taken up by the Romantic thinkers in the early nineteenth century, especially Coleridge. In words that seem to echo both Leibniz and the Stoics Coleridge wrote: 'Life is the one Universal Soul, which, by virtue of the enlivening Breath, and the informing Word, all organised bodies have in common, each after its kind. This therefore, all animals possess, and man as an animal'.

It is, however, in our own time that such ideas are at last gaining wide currency. In the early years of this century, Teilhard de Chardin, a scientist and Jesuit priest, sought to apply the mystical insights of Christianity to our understanding of the natural world. He rejected the dualism of Plato and Descartes, taking instead the ideas of Irenaeus as his starting point. Like Irenaeus he believed that there is in Nature a tendency towards harmony, in which the *logos* present in all creatures draws them into unity one with another. He describes the *logos* as 'psychic energy', comparing it with a magnet which attracts the universe towards the 'omega point' of time, in which the perfect unity of Eden will be restored. The scientific implications of this view are enormous. Scientists, according to Teilhard, should continue to study the process of cause and effect. But they should recognise another process at work, through which the universe is fulfilling its destiny. As Teilhard put it, there are two tendencies within each living creature, working simultaneously: an outward tendency of cause and effect; and an inward tendency towards self-perfection. The role of the scientist, and of the human race as a whole, is actively to promote the harmony latent within the natural order.

Although many scientists feel uneasy about some of Teilhard's terminology, a growing number are expressing similar ideas. The eccentric physicist Capra, seeking inspiration from the Eastern religions, has spoken of the 'divine interconnectedness of all living things'. And even conventional scientists, like Arthur Peacocke, are happy to write that 'the world is, as it were, within God ... with God acting as a mother, giving life and form to a self-creative world within Her own being'. Currently the most popular expression of these insights is James Lovelock's Gaia hypothesis, named after the earth goddess who was worshipped in primitive Greece as the nourisher of plants and young children. *Logos* philosophy, in a host of different forms, is finally catching on.

None of this affects the way scientists work in the laboratory. They must still patiently study the phenomena of Nature, setting up hypotheses which they then test by experiment. The meaning we attach to their discoveries, however, must be radically changed, as must the way in which we apply them. We can speak of benign and malign processes: a benign process promotes natural harmony, a malign process destroys it; a benign process works with the grain of Nature, a malign process works against it. Thus, in developing new technologies we must seek not only efficiency, but also 'benignancy'.

Since all of us must judge and use technology, we must all learn to assess benignancy. And this may lead to a surprising consensus of attitude, especially on the crucial issue of pollution. Those concerned for the natural environment are apt to imagine that all pollution is wrong and should be eradicated; while others, dismissing such a counsel of perfection as unrealistic, can offer no alternative standard of judgement. So the two sides are at loggerheads. In fact some degree of pollution is inevitable: obviously we must produce sewage, but above that almost every human activity has some environmental impact which is potentially harmful. The problem is the scale of the impact. Possibly the greatest environmental threat is the burning of hydrocarbons, which is the primary cause of the greenhouse effect. Up to a certain level this poses no threat, since the resulting carbon dioxide can be converted back to carbon and oxygen through plants—after all, a forest fire, which can be quite natural, is environmentally identical to driving a car. The problem arises when the level of pollution is too great to be absorbed naturally, and the balance of nature is tipped—with all manner of damaging consequences. The same applies to almost every other form of pollution: it is quantity which matters.

We must, therefore, speak of 'optimal' levels of pollution. Up to a point pollution may be regarded as benign, if the technology causing the pollution confers benefits upon humankind—or indeed upon other species. Above that level pollution becomes malign, upsetting the balance of Nature. Of course, there may be doubt and debate amongst scientists about what constitutes the optimum for each type of pollution. But the principle is one on which all can agree—and which, as we shall see later, has important consequences for practical policy.

The image of the perfect society in Western eyes has until now been the smooth-functioning city: an environment wholly controlled by human ingenuity. The diametric opposite of the city is the jungle, a wild environment unaffected by humankind; and when one encounters such an environment, with so many species of plants and animals living in ecological balance, it is tempting to see it as the true image of perfection. But like every other species we must to some degree adapt the environment in which we find ourselves if we are to flourish: that, after all, is the purpose of our superlative brains and agile hands. The recurring image of paradise in the Old Testament—and also in Muslim, Hindu and Buddhist literature—is that of a garden. Gardeners cannot impose their will on the plants in their care, since they must blossom and flourish according to their nature. Yet they lay out their garden according to their own designs, and ensure that each has space to grow, so that they may all have abundant fruit for them to harvest. The skilled gardener thus preserves the balance of nature, but in such a way as to provide for human needs. There could be no better image for the political morality which we must learn in the coming millennium.

Social Enterprise

Adam Smith spoke of 'wealth' as the criterion for a nation's success: the well-being of its members depended, according to Smith, on their ability to produce material goods. The Hebrew people of the Old Testament frequently prayed for prosperity: to them the visible sign of God's blessing on a family was a prosperous home and farm.

On the face of it wealth and prosperity appear synonymous: the wealth of the successful industrialist seems like the prosperity of Abraham in modern form. But on closer inspection they have rather different connotations. Smith's wealth is only a single dimension, a vertical line pointing upward, whose height can be measured in money: becoming wealthy means acquiring financial assets which can be used to purchase material goods. Prosperity, by contrast, has a second, horizontal dimension also. The prosperous patriarch not only has abundant cattle and land, but his cattle shine with health, his soil is fertile—and, more important, his own activities are ensuring that the cattle remain sleek, and the soil maintains and even improves its richness. This prosperity includes the notion of wealth, but it also implies harmony with the natural environment, so that the wealth can be sustained from one generation to the next. Indeed, almost every reference to prosperity in the Old Testament looks forward to future generations.

Today those who are indifferent to, or even contemptuous of, environmental issues often express fear that 'green' policies would condemn us to a dreary poverty: better, they seem to imply, for the human race to enjoy a short, rich life, than a long, poor one. And the green movement has indeed seemed to advocate a drastic drop in living standards, with its exponents showing a puritan pride in their own frugality. Long-standing members of the movement will have sat through many long, intricate debates on personal life-styles, in which those who own cars and large houses stand accused of ecological sin. Such green puritanism is based on a conviction which is profoundly misplaced, that the horizontal dimension of prosperity can only be safe-guarded by denying the vertical dimension: that ecological balance must be achieved at the expense of material wealth. Small wonder that the majority of people have remained stubbornly deaf to the green gospel, despite dire prophecies of ecological disaster.

Green politics can only succeed if it embraces both dimensions of prosperity. This is a matter of scientific and religious truth. The Hebrew tribespeople praised God for the sheer abundance of his creation, marvelling that a single grain planted in the soil could, in a few short months, bring forth a hundred grains; and they believed, as an article of faith, that they were entitled to enjoy this abundance if they were good stewards of the soil.

Over the past four centuries our scientists have discovered that God's creation contains riches beyond the wildest Hebrew dreams: they have uncovered forces and powers within Nature that would have inspired in the Hebrew heart the highest hymns of praise to God. We are equally entitled to those riches—if we too are good stewards.

The challenge of our time may thus be put quite simply. In recent centuries our efforts have been diverted solely to the vertical dimension of prosperity, and in this we have enjoyed astonishing success. Now, as a matter of urgency, we must turn our attention to the horizontal dimension, so that the wealth we now enjoy can be sustained.

The science and technology of ancient Hebrew society was, of course, by our standards very crude. But their

politics were remarkably sophisticated; and for a number
of centuries they seemed to achieve an astonishingly high
level of prosperity, within their technological limits. In
modern terms, their political system appears like a
combination of capitalism and socialism, in which individuals enjoyed a high degree of freedom within a strict
framework of central regulation. But the secret of its
success was that it was profoundly different from both
our modern systems. As we found earlier, both capitalism
and socialism place economic power and social responsibility in the same hands, and so have an inherent
indifference to the environment. The Hebrew system
made a strict division between those who had direct
control over resources and those who ensured that they
were used in the broad social interest. Hence Hebrew
politics was profoundly green.

In Hebrew society land and all other productive
resources were owned by individual families, and they
were free to sell their goods to the highest bidder so long
as they were honest in their dealings—as in modern
capitalism. Thus the individual families worked hard and
efficiently, using their talents and resources to the best
advantage, knowing that they would enjoy the fruits of
their labour. But their freedom was constrained by strict
economic laws, whose purpose was to prevent a family
exploiting either other families or the land in their
charge. Every seventh year the land had to be left fallow
to restore its fertility; and any crops that grew wild had
to be left for the poor to gather. Every fiftieth year any
land bought in that period had to be returned to the

original owners, to prevent successful families from growing too rich while others were pushed into poverty. There were also elaborate rules about lending, the use of crops, slavery, employment of labour and so on. These laws did not stifle individual enterprise; they established a social framework to ensure that the activities of each individual created prosperity for all.

Throughout the period in which these laws were in force, the Hebrew tribes had no chiefs or kings. Instead they had judges, who were usually elderly men renowned for their wisdom. They had no special wealth of their own, and no power to overrule the heads of families. Their only function was to ensure that the law was obeyed. They thus interpreted the law where its application was unclear, settled disputes and punished offenders. This system of judges was unique to Hebrew society; and when they were swept away, to be replaced by a king, the old liberties which individual families had enjoyed were lost forever.

Our society today is, of course, infinitely more complex than that of ancient Israel. Yet the same economics and political principles are equally applicable. Indeed, in the early nineteenth century, in the wake of the Industrial Revolution, a group of political economists led by Jeremy Bentham developed a similar philosophy of government. They recognised that the market system, in which individuals own resources and are free to use them to their best advantage, was the most efficient means of running the economy. But they lacked Adam Smith's faith in an invisible hand that would automatically turn individual self-interest to the greatest social good. Rather they believed that the state had a duty to ensure a 'junction of interest', by enacting laws and regulations to constrain and guide the actions of individuals. They proceeded to undertake detailed research into a wide variety of spheres, such as education, the relief of poverty, health care and waste disposal, to recommend precisely what form the regulations should take. They fervently believed that schools, hospitals, refuse collection and even workhouses should remain in private hands or under the charge of local, independent trusts; the state's role was to set and enforce minimum standards and, in some cases, to provide a subsidy. Unfortunately this approach was not followed, as a division soon appeared between those who saw no economic and social role for the state, and those who wanted the state to manage the economy and welfare services.

Early this century a professor of economics at Cambridge University, A.C. Pigou, who regarded himself as a disciple of Bentham, was the first to recognise pollution as the single major problem facing humankind. He diagnosed the cause as a failure of the market mechanism, whereby firms do not pay the cost which their pollution imposes on the community as a whole. He believed, therefore, that the polluter should be compelled to pay through incurring a tax on the activities equal to the cost of the damage caused. Such a pollution tax would have the effect of reducing pollution to its optimal level. Pigou did, however, perceive a major stumbling block to pollution control: pollution is indifferent to state boundaries. Thus if the atmospheric pollution of our country is blown by the wind across its borders, its neighbours suffer the main effects. There is little incentive for the guilty country to clean up its activities. Pigou, writing in the middle of the First World War, hoped that one of the fruits of peace would be an international agency for pollution control.

Today pollution, above all other issues, is prodding us towards a Benthamite—or rather a Hebrew—philosophy of government. Unfettered capitalism clearly offers no hope of saving the environment, despite the pressure of consumers for more 'eco-friendly' products. And socialism too is utterly discredited. It is clear, therefore, that industry must remain in private hands, but that its activities must be guided and constrained, to achieve a 'junction of interests' between private profit and ecological well-being.

But the global nature of pollution is pushing us beyond Bentham. He, like the ancient Hebrews, could regard all economic and social problems as essentially national, to be solved by national governments. But the nation state is manifestly too small to tackle the major threats to our survival. A single nation on its own can have only a negligible impact on the greenhouse effect or ozone depletion, even if it banned all cars and all CFCs overnight. Equally a country which includes large tracts of rain forest within its borders may be reluctant to lose the profits from cutting down the trees, when it is the world as a whole which benefits from the preservation of the forests. After all, many tropical countries like Brazil desperately need the foreign exchange they earn from selling hardwood timber and raising beef on the acres cleared of trees—not least in order to pay interest on past Western debts.

Pigou's dream of an international agency for the environment is now an urgent necessity. It must enjoy the power, hitherto confined to national governments, of enforcing its regulations. And it must be able to impose taxes, both to control pollution and to send funds to countries, such as Brazil, to help them preserve their natural environment.

Pigou, however, saw a further limitation on the nation state. Some forms of pollution and environmental damage are purely local, without any effect beyond a small area. More importantly, the practical implications of global pollution control will vary greatly from one region to another. So national governments must not only pass power upwards to international agencies, but also downwards to local authorities. Land planning, including the preservation of woodland, is best conducted at local level, as is waste disposal, sewage, transport planning and so on. Indeed as many environmental issues as possible should be entrusted to local government, since public concerns and pressure is likely to be strongest where people can see that their own neighbourhood is at stake.

One of the major stumbling blocks in ecological policy is the extraordinary loyalty, bordering on fanaticism, with which we regard the nation state. We have been brought up to believe that national governments should be sovereign, exercising absolute authority within their borders, and that every ethnic group which regards itself as a nation should enjoy political independence. Thus national governments are profoundly reluctant to cede power to international bodies, or to devolve power to local authorities. Yet even in Europe, which was the birthplace of the nation state, the notion of national sovereignty is little more than 400 or 500 hundred years old. And in other continents it dates back only to the last century when the European powers, in establishing their empires, inadvertently exported nationalism—which in due course fired the imagination of those over whom they ruled, prompting them to demand independence. As a result the continents of the globe are now criss-crossed by national boundaries, many of which are quite arbitrary.

The problem of nationalism was confronted in the most poignant moment in Hebrew history—and that moment, two and a half millennia ago, offers us a lesson now. The Hebrew people demanded a king instead of judges, to make their nation strong and lead them into war. The prophet Samuel pleaded with them, warning that under a king they would never again enjoy the liberty and stability that the ancient law provided: kings might at times inspire them, but they would always tend to oppress them, changing the law to suit their own interests.

Nationalism is always, in effect, rule by kings, even though their guise may change. When a people desire to be a nation, they call for strong rulers to inspire and unite them, and to lead them into battle against their foes. And even when a nation becomes a democracy, they vote for charismatic presidents and prime ministers, who offer simple, cosy solutions to hard, complex problems. But if we were to return to the Hebrew and Benthamite philosophy of government, the age of strong national leaders must close. We need wisdom, not charisma. We want calm courage to enact laws, impose taxes and grant subsidies which command common consent; we no longer require the slick sloganising of power-hungry political parties. Above all, appeals to narrow national sentiment must be rebuffed in favour of a vision which both embraces the whole planet and respects local loyalties.

Such hopes may seem piously idealistic. But changes in political style will follow automatically from changes in political values. Over the past two centuries the false opposition between capitalism and socialism has been reflected in the growth of political parties committed to one or other ideology competing for power. They have naturally thrown up charismatic, nationalistic leaders making lavish promises to the electorate. There is already a strong undercurrent of suspicion of this adversarial form of politics. And the new awareness of environmental issues is now bringing this current to the surface, threatening to sweep away these redundant creeds. There can be little doubt that, within a few years, every party seriously seeking power will be compelled by public opinion to place environmental issues at the top of their list of priorities. And electorates will choose between them, not on ideological grounds, but on the wisdom of their ecological policies and their competence in carrying them out. In Hebrew terms, we shall be asking our politicians to be judges, not kings.

6

Moral Harmony

Spirit of Community

Historians of religion have never ceased to be astounded at the rapid spread of Christianity through the Roman Empire. Within three centuries a small sect, arising from a despised tribal people on the edge of the empire, won the allegiance of a majority of the people in the Mediterranean world. It had the most meagre material resources and suffered continuous persecution, yet it gained such respect that in AD312 Emperor Constantine declared it the official imperial religion, in the hope that it would unify the people under Roman rule. The explanation is simple: as Henry Chadwick has observed, Christians in this period loved one another, not merely in theory, but in practice. When Christians were ill or impoverished, their fellow Christians would sustain them, giving whatever help or money was needed. The Christian Church was thus an extended family based on spiritual rather than blood ties—or, as some Christians saw it, a voluntary tribe. And the theological foundation of these ties was the Christian conviction that soul and body are a single entity, so that spiritual goodness within the soul must find outward expression in acts of material generosity.

As Paul himself had observed, the belief in the unity of body and soul was a scandal to the Greeks, raised as they were on the philosophy of Plato. And, as we have seen, Platonist ideas, in somewhat muted form, had successfully undermined the original gospel by the fourth and fifth centuries. Indeed once Christianity had become the established religion, the revolutionary love of the early Christians was regarded as dangerous: far better to allow the Church to concentrate on purely private spiritual matters, leaving the emperor a free hand over public, material affairs. Thus Platonist dualism became embodied in the strict division of function between Church and state.

But just as there was continuing philosophical resistance to this dualism, as we saw in the last chapter, so groups of men and women created communities that sought to emulate the practical love of the early Church. As the Church under imperial approval grew increasingly wealthy, tens of thousands trekked out into the deserts of Africa and the Middle East to set up monasteries, in which every aspect of life could be moulded according to the gospel. The monastic pioneers saw the communion service, instituted by Christ himself, as the symbolic expression of their vision. The bread and wine represents the fruits of human labour: thus, in consecrating them on the altar, all labour becomes holy; and, in distributing this equally to everyone, the community expresses its commitment to economic justice for all.

To a monk like John Cassian, whose writings were widely influential, the monastic movement was political as well as religious in its aims.

Not surprisingly the monasteries were often regarded with grave suspicion by bishops and priests, as well as by emperors and kings, who rightly regarded them as providing a moral threat to their privileges. For a millennium they remained the spiritual power-house of the Church, but as their wealth grew they too became corrupt. However monasticism demonstrated an extraordinary capacity for self-renewal, every century or two rediscovering its original vision. And as the tribal family system collapsed across Europe, so the monasteries became centres of education, and of care for the sick and destitute. Indeed the wealth of the monasteries was largely founded on this social role, since people willingly contributed to them to support their work.

The Protestant Reformation in the sixteenth century swept away the monasteries, at least in northern Europe, and their social role was to a great extent taken on by charitable trusts. But the true inheritors of the monastic vision were the Christian Socialists, who flourished throughout Europe and North America in the aftermath of the Industrial Revolution. Many looked back to the monasteries for inspiration, and sought to interpret their vision within the new industrial cities. Despite their name, they were far from being socialist in the normal sense of the term. They wanted the people themselves to come together within their neighbourhoods to form schools and hospitals to meet their own needs. In this way, they believed, the 'voluntary tribe' of the early church could be recreated amidst the grimy tenements and terraces of the industrial west.

Their most formidable philosopher and theologian was a shy Anglican priest, F.D. Maurice. Starting from the concept of *logos* in John's gospel, he followed Leibniz and Coleridge in regarding the whole universe as 'sacramental': beneath the diversity of the material creation there is spiritual unity. But his genius lay in applying this insight to human community. Human relationships, Maurice believed, cannot exist in an economic vacuum, but must be founded on common work and common material interests: it is through working together, and through sharing practical decisions, that we grow together in love and friendship. And as we grow together spiritually, we will naturally want to express our love in economic justice, in which the fruits of our common labour are distributed fairly.

To Maurice, therefore, the fundamental sin of modern industrial society is that it destroys human community by creating a division between the spiritual and material aspects of our lives. The nuclear family, in which husband and wife are together only during the brief hours of leisure, was to Maurice a moral monstrosity since it deprived the family of the material bonds which sustain human relationships. And, outside the home, men and women are 'mere pebbles on a beach, physically close but spiritually isolated'. In the factory or the shop 'people exchange their labour or goods for cash, while their souls barely meet'. Although Maurice and

Marx could never have read each other's writings, since they wrote in different languages at almost the same time, Maurice's spiritual analysis of capitalism echoed precisely Marx's theory of alienation.

Maurice's first practical campaign was to start small industrial co-operatives, where workers owned the businesses and shared the decision making. Sadly most of these co-operatives failed, partly because the workers lacked managerial expertise and partly because they lacked capital. He then turned to education, starting a college where working men and women could study in the evenings and on Sundays. This proved far more successful, inspiring similar projects all over Britain and then further afield. Maurice concluded that it is in the sphere of social welfare—education and health care in particular—that the renewal of human community must begin. In pre-industrial societies the nurture of the young and the care of the sick are the primary roles of the extended family, and thus the basic building blocks of community. By depriving the family of these roles, industrial capitalism was undermining family life and throwing people into spiritual isolation. Maurice saw no possibility of restoring the extended family within our modern cities. Instead he advocated neighbourhood co-operatives in which groups of families should get together to provide education and health care for themselves.

Although Maurice's writings were never widely read—his prose is almost impenetrable in its verbose complexity—his ideas kept bubbling up, with many different people claiming to have been inspired by him. At one remove, via John Ruskin, he influenced the greatest modern prophet, Mahatma Gandhi. Remembered now for his campaign for Indian independence, Gandhi himself saw his primary mission as the renewal of village life in his country. He perceived that trade with the West was turning Indian villages from self-reliant communities into slave-camps, forced to obey the vagaries of world markets. He therefore advocated village-based co-operatives producing all their basic needs, and village trusts to run schools and clinics. And the heart of his vision was the conviction that human beings can only flourish if the material and spiritual aspects of their lives are unified. Out in the field and in the workshop this means caring for the soil and using local, renewable materials; at home, in the school or in the clinic, it means forging personal bonds, with people striving together for the good of all.

Gandhi also gave his attention to the methods of education and health care themselves. He believed that the child should not merely be regarded as a mind to be filled, but also as a soul to be nurtured. He was thus sceptical of the Western approach in which children are confined to a special building for the entire day, where a single teacher faces a large number of pupils. Instead he preferred the traditional Hindu method, in which the guru meets his pupils in small groups, for only one or two hours each day, discussing their progress and setting them exercises. He recognised that this would require more teachers than a Western-style school, and that in financial terms it would therefore be more costly. But, as he observed, any system which regards time spent with our children as a cost to be minimised, rather than as a gift to be lavishly bestowed, is profoundly immoral.

Similarly the sick patient should not merely be seen as a body to be repaired, but as a living spirit to be made whole. Thus his scepticism about the large classroom applied equally to the crowded clinic and hospital ward. Just as education requires a close personal relationship between teacher and pupil, so healing requires closeness between physician and patient, so that the physician can understand the spiritual, as well as the physical, dimension of the disease. Gandhi believed that such an approach would require far fewer pills and potions, since the physician's task would be to harness the self-healing powers of the patient.

Today the insights of Maurice and Gandhi, and others writing in similar vein, are becoming so widely accepted as to require little further argument. In the field of health care even the most sceptical doctors are acknowledging the wisdom of a more holistic style of medicine in which spirit, mind and body are treated as a single entity. Teachers too now largely accept a holistic approach to education, rejoicing in the spiritual insights of their pupils as well as their intellectual dexterity. Even in industry managers have come to see the virtues of people working in small groups and sharing decision-making: as Schumacher observed in *Small is Beautiful*, when an emotional commitment to the firm's activities is created, the workers' productivity also rises.

But if attitudes are changing rapidly, institutions remain stuck in the economic mud. Doctors are faced with the unenviable choice of remaining with public health care, where they are chronically short of funds, or going into private practice where they treat only the affluent minority. Similarly teachers, who might wistfully wonder whether there are more sensitive and humane ways of preparing children for adulthood, know that even in private schools the large classroom is the universal forum for education. Indeed, quite apart from financial constraints, parents in Western society would recoil in horror at Gandhi's approach, where the children would only be with their teacher for a short period of the day: our frenetic urban lives leave no room for children for the rest of the day.

Thus the movement towards holistic health care and education requires not only a change in philosophy, but a fundamental reappraisal of our attitudes to work, resources and family life. In short, we need a fresh understanding of the human cycle, from cradle to grave.

Social Nurture

To be a growing child in a typical tribal society is to move outwards through a series of concentric circles. At the centre is the mother within the family home, and during the first months she alone tends the child. Soon a wider circle of adults is discovered, starting with the father, and then uncles and aunts and grown-up cousins, plus grandparents and other members of their generation, all living nearby and intimately concerned with the child's well-being. At first the child is simply an onlooker, carried round on the mother's back or hip, watching the daily activities of the village, bouncing up and down as mother sows seeds, grinds grain and sweeps the home. Once the first faltering steps are taken, the child begins to be taught the wide variety of skills needed for adulthood, from climbing trees to gather fruit to sharpening sticks for hunting. Then as puberty approaches the child may also go to tribal priests for instruction in religious ritual, and to learn the sacred texts by heart—in ancient Israel the boys were taught the Torah, the laws governing every aspect of daily life. The child may then for a period of months attend a school, along with children from other families, to be taught the values of the tribe by priests and elders. Finally there is a solemn ritual to make the transition from childhood to adulthood.

Thus in the tribal culture, adult activities allow ample space for children, who can from an early age make a useful contribution to the family economy. At no stage has the child to make a leap from the intimate mutual care of the family to a harsh competitive world outside: since society itself is based on the extended family, the skills and attitudes which are learnt within the home are deepened and refined as the child grows towards adulthood.

At the other end of the human cycle the elderly too are useful and valued members of the tribal group. In many tribes it is the old who wield political authority, while the young concentrate on economic activities: through their wisdom and experience the elderly are better able to see both sides of a dispute and to judge between them, making prudent rather than impetuous decisions. Also as their physical powers fade, the elderly can give more time to the education of children: in Israel groups of children gathered at the village well to hear the old men relate the history of the Hebrew people. There is no sudden break, after which the person is deemed past work; rather there is a gradual change, during which the balance of a person's work shifts from active to sedentary occupations.

Similarly there is no clear separation between health and sickness. The Hebrew laws, in common with the customs of most tribal societies, give almost no explicit guidance for the care of the sick. Rather periods of sickness, during which a person's powers are diminished, are regarded as part of the natural course of life. During such times a person simply does less work and requires greater care until health is restored. Early anthropologists saw many of the rituals which priests and shamans performed for a sick person as forms of magic, which people falsely believed would bring healing. Closer studies have revealed a rather different interpretation. The curing of illness is usually ascribed to Nature, and to the latent divine power within it; the purpose of the rituals is to awaken and harness those powers, by inspiring within the sick person the will to survive.

The amount of time spent in most tribes on education and the care of the sick far exceeds that spent in modern

Western society. In material goods we are rich, but in social welfare we are very poor. The reason is simple: in our society social welfare incurs a financial cost, paid for via taxation; whereas amongst tribal people it is part of the normal life of the family, as natural to them as watching television or reading a book is to us. And one of the additional benefits of such a system is that divorce and family breakdown are virtually unknown and unthinkable: families are bound together in a tight web of mutual care.

It would, of course, be impossible to return to a tribal form of education and health care. And it would be undesirable too. Despite its many virtues, tribal education is narrow and intellectually unstimulating; in our complex culture, both a broader and a more cerebral education is required to prepare children for adulthood. And in the care of the sick we should rejoice in the technical achievements of modern medicine, even while we doubt the unduly mechanical means with which they are applied. But we can derive two vital, and closely connected, lessons from the tribal example. First, education and health care should somehow be integrated with the whole life of the family; and secondly, the relationship between teacher and child, physician and patient, should be based not merely on a professional contract, but also on mutual personal commitment.

Although it had long been declining, the rapid collapse of family-based education and health care occurred during the Industrial Revolution as people moved from village to city. The extended family was destroyed at a stroke, and husband and wife working long hours in a factory had neither time nor space for any other activities. Yet the memory of life in the village continued to stir the urban imagination. And during the nineteenth century throughout Europe and North America ordinary working people set up friendly societies, co-operatives and trusts to educate their children and tend their sick. Sometimes rich benefactors would provide finance, but to a quite remarkable extent these ventures were funded and managed by the workers themselves. Thus, for example, by 1870 85 per cent of all children in Britain received elementary education, mostly in small school-rooms established and run by their parents.

Most of those local schools were taken over by the state or else swept away, as were the various other local welfare institutions. Yet their existence in the midst of the industrial slums was an extraordinary testimony to people's ability to provide for their own social needs. And today, as the state welfare system is condemned to a slow and ugly death, the example of our nineteenth century forebears is a lesson for the future.

In the early nineteenth century, before the idea of a welfare state had been mooted, many political economists welcomed the local trusts and co-operatives. They believed that competition between independent schools and hospitals would encourage a high standard of service. And, since the local people who used and funded them also managed them, there was a strong incentive to be efficient. They recognised, however, that the effective-ness of local trusts would be inhibited by the poverty of working people, who could afford only the smallest contributions. They thus wondered how best the state should help. All sorts of different schemes were suggested, from the tax rebates proposed by Tom Paine, to a system of vouchers outlined by Nassau Senior. But on one vital point all were agreed: that the state should subsidise the users, not the institutions. So the young and the sick should receive help in paying their fees, rather than the schools and hospitals receiving direct subsidies. This would ensure that the schools and hospitals remained under local control, and subject to local competition.

Sadly such wise advice was rejected. As early as 1833 the British parliament was voting subsidies to certain educational societies, and within half a century the state had taken over the entire system. But there is no reason why the process could not be put into reverse: indeed there are already various cautious measures, in Britain and elsewhere, in this direction. Although there are innumerable administrative hazards, there are broadly two steps. First individual schools and colleges, clinics and hospitals would be taken out of state control and handed over to voluntary trusts, attracting state finances according to the amount of work they do; there is thus local control plus an incentive to operate efficiently. Secondly state funds would in some form be allocated to individuals rather than institutions, according to their means: richer people would then pay from their own pockets, thus allowing the total amount spent on education and social welfare to grow.

Once these steps had been taken, the way would be open for a gradual change in the methods of education and health care, according to the preferences of individual families. Some, no doubt, would continue to want their children to attend large schools, happily paying a full fee in return for having the child taken entirely off their hands. Others, however, might prefer other styles of education; and people would be free to establish their own trusts and co-operatives, as they wished. Thus one imagines, for example, the development of academies along the lines of those in ancient Greece and India, where the teacher saw pupils in small groups for short periods, setting exercises for them to do, and parents were responsible for ensuring that their children performed these exercises. Also small schools might spring up, in which parents themselves administered and maintained the school, employing a small number of professional teachers. None of these ideas is unduly fanciful; they have flourished in the past when people were free to educate their children as they pleased. Indeed such freedom releases enormous creative energy, enabling imaginative individuals and groups to set up schools according to their own vision.

There would be similar variety in health care—as is already occurring amongst 'alternative' practitioners. Since many of those in hospital require merely nursing, with comparatively little expert treatment, smaller neighbourhood hospitals would be likely to prove more popular than the huge structures that have become fashionable under the state system. These would, one imagines, continue to be complemented by a small number of specialist units for more complex care. It is clear that many people are willing to pay more f᠈. a local doctor who can spend time listening to the patients and responding in depth to their condition; so the number of doctors would multiply. And, of course, there would no longer be any distinction between orthodox and alternative therapies, since every form of treatment could flourish according to people's perception of its value. In addition, as with schools, many people would willingly commit themselves to the management and maintenance of hospitals and clinics, perhaps enjoying reduced fees in proportion to the work they do.

Under such a system the state would retain three vital roles. First, it would need to set basic standards in both education and health care, and employ inspectors to enforce them. Ordinary people are unable to make a full judgement of the quality of these services, because both involve a level of professional expertise which the lay person cannot assess—the person on the point of having an operation is in no position to weigh up the skill of the surgeon. Thus the state would have to appoint experts who could make that judgement on our behalf. However it is important that in setting standards, the state should exert only the minimum amount of control, allowing the institutions the greatest possible freedom to develop their own styles and methods.

The second role of the state would be to subsidise the poor. At present, by financing the institutions themselves, the state is in effect paying equal amounts for everyone, regardless of ability to pay. This is a ludi-crously inefficient form of state expenditure. Instead subsidies should be paid to pupils and patients according to need. Since this would greatly reduce the total state expenditure on welfare, it would be possible to reduce taxation. Thus richer families would not be significantly worse off, since their payments for education and health care would be offset by lower tax bills.

The state's third role would be to supervise welfare insurance—a role that was first envisaged by Adam Smith himself. Obviously the costs of education are concentrated in a particular phase of life. For all but the richest families these costs need to be spread out over a longer period, through payments into an insurance fund. Health costs are highly unpredictable, both in size and in timing, and again insurance funds are the only means of overcoming this risk. As with education and health care themselves, insurance involves a degree of expertise beyond the grasp of most of us. Thus the the state would have to inspect and approve insurance companies, and offer a guarantee against default. Indeed the state should compel everyone to take out insurance, just as motorists are currently required to do, to prevent destitution.

There is little doubt that freeing social welfare from state control, and allowing people to provide and pay for it themselves, would stimulate a huge expansion. In the jargon of economists, education and health care have a very high 'income elasticity of demand'—people wish to spend an increasing proportion of their income on them as they get richer. Yet under the state system the demand is frustrated, since constraints on tax revenue put a constant pressure on schools and hospitals to cut expenditure. Once that constraint is lifted, teachers and lecturers, doctors and nurses, would be astonished at their good fortune.

And, if the nineteenth century experience is an indicator, the poorer groups would share in this growth. Released from their position as helpless dependants compelled to accept whatever the state provides, the poor would be free to improve their own lot by their own efforts. History teaches that such opportunities are eagerly seized, so that one can imagine schools in poor areas achieving standards at least as high as those in richer districts, thanks to the time and zeal expended by parents—and grandparents!

It is hard to predict the impact of such changes on family life. It is unlikely, and perhaps undesirable, that the extended family would reappear: the population is likely to remain too mobile, and families' ties would therefore remain too fragile. On the other hand married couples who were helping to run their school and clinic would enjoy a shared practical interest; and they would find themselves part of a wider network of personal relationships centred on these institutions. Thus the isolation of the family would in part be broken. This could only help to reduce the rot of marital breakdown—even if the change were slow and undramatic.

PART FOUR

WHOLENESS

7

Holistic Economy

Prophetic Politics

The Old Testament is sometimes described as 'salvation history'—the history of how the Hebrew tribes sought freedom and justice. Far from being a smooth progress, it is punctuated by defeats and catastrophes, and ends seemingly on the brink of total failure. The reader can discern strong movements and trends, which at first sight seem irresistible, as if the course of history were predestined. But at moments of acute crisis prophets appear, spelling out the moral choices which the people and their leaders face; and when the prophet's voice is heard, the current of history can change. Indeed the word 'crisis' derives from the Bible, meaning 'judgement'. As the Hebrew prophets constantly reiterated, human destiny depends on how we respond to God's judgement, made manifest in history itself.

Today it is common to describe our condition as one of crisis, the most dire crisis that humankind has faced, because the very survival of our species is at stake. To many people, ultimate disaster seems inevitable: human civilisation is like a vast ocean liner heading towards the rocks, too late to alter her course. One cannot help wondering if the spendthrift ways of the people of Europe and North America are a final fling before the liner sinks: opinion polls over recent years confirm that most people expect the worst, and respond by reckless over-consumption.

Yet prophets can change the course of history. Amidst the storm, the prophetic eye can discern the currents flowing in the right direction, away from the rocks. The prophetic voice, if heard, can show us how to catch these currents. And one need not go back to ancient Israel to see the prophet at work. It is difficult to imagine the flowering of industrial capitalism without the strange Scottish prophet to whom, somewhat perversely, this book is dedicated—Adam Smith. When he wrote *The Wealth of Nations* Britain and the rest of Europe, seemed to be stuck in an economic and political bog: high unemployment and low wages, combined with the collapse of imperial pretensions in the New World, were pulling the civilised world down towards revolutionary chaos. But Adam Smith had visited small factories in the Pennine villages and on the banks of the Clyde, and saw in them the hope for the future. His genius lay in describing an economic and political system that would turn those factories into vast industries, shipping their products across the world.

Are there today green equivalents of those early factories, that can become beacons of hope for the third millennium? Are there benign currents that we can catch, which will carry us into calm and peaceful waters? In the past two decades four new trends have appeared which viewed separately seem insignificant, but taken together possess great power for good.

The first is the shrinking of the factory, caused by the rapid development of microelectronics. Already computers have caused great changes in work-patterns, eliminating many humdrum clerical jobs. But these are minor compared with the potential revolution in manufacturing—of which as yet we are seeing only the early signs. The old industrial machine could do a single simple task very rapidly: thus a factory required a long series of such machines, housed within a vast building. The 'intelligent' machine in contrast, can perform many tasks in sequence, so that far fewer machines are involved. Additionally, while old machines could not be adapted to different tasks, the new machine can quickly be reprogrammed. The implications of this technological leap are enormous. Instead of sprawling cities built around huge factories, manufacturing can flourish in market towns and even villages, in relatively small workshops. Instead of the market being flooded with masses of identical products, firms will be able to produce far smaller runs of each item, adapting it to the needs of the customers; and this in turn will put the premium on quality rather than quantity. Instead of one area of the world—be it Europe last century, or East Asia today—dominating the global economy, firms will find it more profitable to put small factories in every country, where they can respond to the local market. Today this revolution is rapidly gathering force: over the next half-century it will have an economic and social impact as great as that of industrial capitalism in the fifty years after Adam Smith's great work.

The second trend is closely related to the first: the dispersal of population. For at least two centuries, until around 1970, people were moving from the countryside to the cities; now the population is shifting back towards small towns and villages. In part this is due to faster roads which enable people to commute more easily into the cities for work—and this, of course, is ecologically malign. But now work is following the people. Offices and workshops are moving out of the cities, and growing numbers are able to work from home. As with the first trend, the key is technological: huge improvements in electronic communication. Today an office can be in a remote mountain village and still be in as close touch with its markets as if it were in a city—so long as a fibre-optic cable snakes its way up the hillside.

The third trend is in agriculture: the return to mixed farming is ceasing to be the eco-freak's dream, and is becoming an economic necessity. Until recently anyone criticising modern farm practices was challenged by statistics showing huge increases in yields per acre. Peppering the soil with chemicals, it was held, saved the world from starvation. Indeed, in a strange irony of terminology, the introduction of agrochemicals, and then of high-yielding crops, in India, Africa and China, was hailed as a 'green revolution'. Now yields are starting to drop. In many parts of Western Europe and North America over a tenth of the farm land is bare of top-soil, turned to barren dust by chemicals and heavy machines. We are now suffering the fate of every civilisation that has tried to squeeze more from the land than the soil can give: we are turning our fields into deserts. Today, however, there are no fresh pastures to discover, no under-populated continents to colonise. Instead farmers are beginning to rediscover that the land will only sustain its yield if it is fed as well as tilled. And the judicious mixture of livestock and arable agriculture uses nature's own method of restoring fertility. As with the technological revolution, this too has profound implications. Mixed, organic farming requires more labour and can be conducted on a far smaller scale than chemical-based agriculture. At a time when many people are craving more physical exercise, once can imagine people happily working part time on the land, as a complement to their effort at a computer terminal.

The fourth important trend, which we have already explored, is within social welfare. Industrialisation in the nineteenth century saw a sharp drop in the time which adults devoted to education and health care, simply because both men and women were away from home for most of the day—and it had been the home where the young had been taught and the sick tended. But over the past three or four decades, in response to popular demand, there has been a huge rise in expenditure in these spheres, mainly via taxation: as people have grown more affluent, social welfare has gained an increasingly high priority. As we have seen, this rise is now largely frustrated by an absolute limit on the amount that can be raised through taxes. Nonetheless, the demand is, as it were, continuing to build up like water behind a dam. If the demand could be released, social welfare would without doubt be the fastest growing sector of employment.

There are, of course, experts in different academic disciplines charting each of these four trends: economists studying the shrinking factory, geographers looking at the dispersal of population, agriculturalists predicting a return to the small mixed farm, and sociologists puzzling over the paradoxes within the welfare system. But, precisely because these trends are rarely perceived by a single eye, we have until now missed their revolutionary power. Taken together they point to a social and economic transformation as profound as the Industrial Revolution two centuries ago—a transformation which offers hope to our rotting civilisation.

Predicting the future is, of course, notoriously hazardous. But the prophetic eye does not so much foresee the future as perceive possibilities that are within our grasp. And these trends allow us to prophesy a pattern of life for our grandchildren and their offspring that is both ecologically sustainable and emotionally fulfilling.

The huge, dirty factory will be virtually extinct, and so the sprawling conurbation will be redundant. Instead the majority of people will choose to live in smaller, more humane communities. Villages and small towns, which have seen their populations dwindle since the Industrial Revolution, will double, triple and quadruple in size. The new houses that will be built to accommodate this great influx will be designed to minimise energy use, with thick insulated walls and the main windows facing southwards. The cities themselves will shrink, reverting to their old role as trading and cultural centres.

Workshops and offices will spring up in the expanding towns and villages, so most people will work near their homes, in many cases within easy walking or cycling distance. Large numbers will work within their own homes, linked to the outside world by an electronic terminal. Thus the number of cars on the roads will decline steeply. Moreover, since the dispersal of population will allow public transport services between village and town to flourish once more, many families will do without the expense and anxiety of running their own cars. And, since goods will be manufactured near their markets, adapted to local needs and wants, the volume of lorries will also decline.

Technological progress will have reduced to only a few per cent the proportion of labour needed to produce the goods we consume. This will enable far more effort to be expended in the design of those goods, so that quality and durability will be greatly enhanced. In addition many people will wish to work shorter hours in the workshop and office, in order to devote a few hours each day to the land. This will in most cases simply mean growing fruit and vegetables in the garden, but a few will operate small farms, either on their own or in co-operation with others, raising livestock and growing grain for the market. Since they will have earned their main wage elsewhere, there will be little pressure on labour productivity, so they will have time to tend the land organically.

These expanding villages and towns will prove to be fertile ground for neighbourhood trusts for education and health care. There will be schools offering full-time supervision of children, in return for a fee. There will be academies where tutors will guide the children's studies with the active involvement of parents. And there will be informal groups of parents teaching their own children. There will be clinics offering general medical care, augmented by specialists, many with surgeries in their own homes, offering a variety of therapies. There will be small local hospitals for people needing constant nursing. And the health- care trusts will contribute to large city hospitals to which they can refer patients needing more complex treatment.

Such a society will obviously use far less of the world's resources, and do far less damage to the environment. Since the quality and durability of goods will be higher, fewer goods will be made and purchased, and the demand for raw materials will be low. Less mobility and greater energy conservation will both reduce consumption of fuel and lower the emission of harmful gases into the atmosphere. And in such a society, in a series of overlapping communities—school, clinic, office, neighbourhood, farm—people will be bound together by a common economic and social interest.

It is a beguiling picture; and, it is tempting for the hard-headed realist to dismiss it as a utopian fantasy. Yet precisely because it is so attractive, it is also feasible. Too often the images of a sustainable future offered by green-minded people have seemed dreary and dark, overcast by a grey cloud of puritanical zeal, and no electorate would willingly choose such a future. But the more appealing a green future appears, the more willing we shall be to make the necessary changes to realise it.

This green transformation will not take place automatically, however. Those who profess hard-headed realism are often opponents of change of any kind, anxious to retain present methods of manufacturing and agriculture and committed to the welfare state. Yet such 'realism' is in fact the height of unreality. Changes, of a most profound nature, are bound to occur. The ecological consequences of our present actions will, if unchecked, have a devastating impact on our climate and landscape, from which none can escape. The question is whether we can, in the near future, enact policies which enhance and hasten the benign trends which already exist in our society. If we can, we will not merely sail past the ecological and social catastrophe to which we are now heading; but we shall attain a new prosperity, in which we can live in harmony both with ourselves and with the natural order.

Practical Action

Those of us within the green movement have, since the early 1970s, bombarded each other with green manifestos. They are often the fruit of fine scientific research, presenting the ecological problems with admirable clarity and offering technical solutions. As a result the public is remarkably aware of the kind of technical and scientific challenges which our civilisation must meet if our planet is to survive. Indeed, none of us would have dared to dream twenty years ago of the extent to which ordinary people have come to understand the causes and cures of acid rain, ozone depletion and the greenhouse effect. This is the major achievement of the green movement to date.

But recently the movement has, with a degree of justice, been accused of a kind of creeping fascism. It seems as if green politics consists of a massive list of rules and regulations, governing almost every sphere of life. The list of ecological recommendations is now so vast and complex that only a dictatorship could enforce them. This is not because green-minded people are naturally authoritarian; on the contrary, most of them treasure personal freedom and, as any organiser of a green event can testify, they are inclined towards anarchy! The problem is that too little attention has been given to the political and moral aspects of our crisis, as distinct from the scientific and technical aspects. The Industrial Revolution needed both its scientists, such as Francis Bacon, Isaac Newton and James Watt, and its political economists, such as Adam Smith and his followers. The green revolution still lacks a coherent political economy—or to coin a more appropriate term, 'political morality'.

Important benign trends are already firmly established in our society, and our first priority must be to sweep away the laws and institutions which constrain them. In other words, we must begin by abolishing rules and regulations, and enhancing individual freedom. Within most Western countries there are four main areas where the state inhibits benign change: land, planning, welfare provision and agriculture.

In the past half-century Western governments have imposed stringent controls on building. Their original purpose was to prevent the continuing sprawl of the major cities, and to preserve the countryside from unsightly development in every pretty village and pasture. Now, however, these controls are preventing the dispersal of people and industry to villages and towns, thus in effect turning the cities into prisons. Indeed in many countries these effects are even more malign. Land is released for housing in and around villages, while offices and factories are kept within the cities. As a result only the rich can afford to move to the villages, commuting long distances each day to work. Thus the volume of cars on the roads increases while the gulf between rich and poor is widened—the rich inhabiting spacious houses in picturesque villages, while the poor remain in densely packed urban estates.

Planning controls must be relaxed to permit the gradual dispersal of both people and jobs. It is clearly wrong to allow houses and offices wherever people wish to build them, since the danger of unbridled development which led to the controls being imposed still remains. But existing villages and towns should be allowed to expand, ensuring a balance between residential and commercial development, so that the newcomers can work where they live. Such a policy will meet bitter resistance from the present villagers, who will plead that their ancient communities will be destroyed. Since the old communities were destroyed long ago, when the village bakers, farriers and wheelwrights went out of business, such pleas amount largely to the preservation of present privilege—and so should not be taken too seriously. Indeed restoring jobs to villages and small towns is a *return* to the ancient pattern.

While the welfare state remains intact, albeit eroded by constant financial cuts, neighbourhood trusts cannot flourish. Any individual or group wishing to opt out of the welfare state is compelled to pay twice over, for state welfare through taxation, and for private welfare through direct payments. Thus private education and health care must remain the preserve of the rich. If, therefore, neighbourhood trusts are to spring up, the welfare state must gradually be dismantled. This is not a question of abolishing existing welfare institutions, but of turning them into voluntary trusts and then allowing new trusts to grow alongside them, with individuals free to choose between them. Since the pent-up demand for health care and education greatly exceeds present supply under the state system, the dismantling of the welfare state would not imply that present schools and hospitals would be destroyed by new competition: there is room for old and new. But the old institutions would undoubtedly be compelled to adapt, responding directly to the wishes of those who use their services.

During the twentieth century almost all Western

governments have spent huge sums of public money encouraging bad agriculture. Their professed aim has been to stimulate higher yields, and to this end they have paid farmers to rip up hedges, cut down woodland, drain natural marshes, plough up ancient meadows and invest in heavy machinery. Worse still, they have raised the price of the crops themselves, stimulating farmers to apply huge quantities of artificial fertilisers, pesticides and herbicides to their soil, in the knowledge that the costs of these chemicals will be recouped. The main beneficiaries of these policies are the landowners, who see the value of their holdings multiply.

The results, of course, have been ecologically disastrous. Left to their own devices, farmers would prove far better stewards of the land. The state has only one beneficent role in agriculture, hallowed by ancient history: to operate a buffer stock, as Joseph did under the pharaohs, and as more enlightened states do today. Since harvests vary according to the weather, the state should buy up surplus crops in good years, storing them to sell in years when the harvest is poor.

Thus great progress can be made towards a green future merely by abolishing malign government intervention. This, however, is not enough. As we have discussed in earlier chapters, the free market is indifferent to the natural environment and also to the human need for community. The role of government, therefore, is to erect a firm and stable social framework, not to stifle individual enterprise, but to guide it for the good of all. This framework has two pillars.

The first pillar is taxation—or, more precisely, taxes and subsidies. The most glaring failure of the market mechanism is that polluters incur no costs for the damage they cause; so firms, compelled by competitive forces to keep financial costs to a minimum, cannot afford to control their pollution—even if they wished to. The state therefore must tax pollution, setting the rate at whatever is necessary to reduce pollution to its optimal level. This in turn will stimulate firms to invest in new methods of production which will actually minimise pollution, since this will reduce their tax bill and raise their profits. The same principle can be applied to agriculture and mining. Agricultural chemicals should be taxed, so that organic methods are financially more attractive; and this in turn would stimulate research into low-chemical agriculture so that yields could be maintained without toxic sprays. Similarly non-renewable resources should be taxed, raising the price of goods which use metals and other scarce natural resources; this would make it more attractive to buy more durable goods, and to repair rather than discard the goods we already own.

Taxation should also be used to curb consumers' pollution, in particular the use of fossil fuels. People are understandably frightened at the possibility of a massive rise in petrol prices, imagining that their living standards would fall drastically. However, the experience in the early 1970s, when oil prices quintupled, should dispel such fears. Car manufacturers very soon improved fuel efficiency by over 30 per cent, while firms sprung up to insulate houses and offices; and if prices had continued to rise, consumers would soon have demanded better trains and buses. Sadly by the 1980s the real price of oil was falling rapidly, so the achievements of the 1970s were largely wasted. But if governments committed themselves to a steady rise in fuel taxes over the next ten or twenty years, our consumption would steadily fall while our living standards would actually rise. After all, journey speeds are actually falling in most Western countries because of road congestion; if instead we turned to public transport, both speed and comfort would improve.

Pollution taxes, while saving the environment, would also boost the government revenues. And these in turn could be used to pay subsidies to people with lower incomes for education and health care. If these services are to be taken out of state control, the majority of people must not find themselves worse off. Thus their payments to private insurance companies for education and health care must be offset by a substantial reduction in taxation. The income from pollution taxes would enable this to happen.

Taxes and subsidies are the most effective means of guiding people's actions. They do not stifle individual enterprise, but rather steer it in the most socially beneficial direction. They thus enable huge social and economic changes to occur, without interfering with personal liberty. However, to some degree and in some areas direct regulation is essential.

The most important area for direct regulation is cumulative pollution—forms of pollution where nature has no means of self-correction. The most obvious example is nuclear radiation: once radiation has been leaked, it remains in the atmosphere or the sea for thousands of years. Another example is CFCs where they build up in the upper atmosphere, taking centuries to disperse. In such cases there is no optimal level of pollution: every extra amount of pollution causes permanent damage to the environment. The only solution is stringent controls to prevent all such emissions.

The second vital area for direct regulation, as explained earlier, is in setting and enforcing basic standards in education and health care. The patient is in no position to judge the competence of a doctor, nor the standards of hygiene in a hospital or clinic. Equally parents cannot be adequate judges of teachers. Thus the state, or legally authorised professional bodies, must judge the skills of doctors and teachers, and the quality of the institutions in which they work. This in turn requires the state to set minimum standards to which all must adhere. There is nothing strange in such a role: it already applies to other professional groups such as lawyers, architects and vets, who practise privately but must submit to strict standards of quality. However, with education and health the state must take particular care not to stifle the less orthodox methods and approaches; so the standards should be set as broadly as possible, while still reassuring patients and parents that they are in safe hands.

There are, of course, many other areas where the state has a vital role, and where in most countries it is already active. It should regulate monopolies, supervise the provision of basic utilities such as water and electricity, set standards for sewage and waste disposal, supervise the broadcasting media, and so on. But in every case the same principle applies. The government, as far as possible, should not attempt to supply the services, but should leave this in private hands; its task is to guide and encourage, through a mixture of taxes, subsidies and direct regulation.

This philosophy of government is neither capitalist nor socialist, nor is it specifically green. Historically it derives from Hebrew practice, and has in various ways been reactivated in the nineteenth century by such diverse figures as Jeremy Bentham and F.D. Maurice. It shares capitalist faith in individual enterprise, while recognising with socialism that state action is essential for social and economic harmony. More importantly, it offers the only realistic means through which green ideals can be put into practice.

8

Holistic Spirit

Prophetic Faith

To some people faced with the present ecological and social crisis, religion may seem a luxury, with little practical relevance. Religious leaders seem to be absorbed in arcane points of theological doctrine, while the majority of people remain resolutely indifferent to all formal worship. Worse still, Western society contains so many different denominations and faiths that religion seems to divide people rather than unite them. It can no longer rally people to a common cause, as it has so often done in the past.

Yet religious symbols and attitudes remain powerful. The social philosopher Emile Durkheim, though himself an atheist, believed that religion is the means through which moral and social values are transmitted and maintained. So even when few people profess any religious creed, they will still cling to the old vestiges of faith: church buildings, Christmas carols, harvest festivals, remembrance parades. And in this century, the figure who, more than any other, caught the world's imagination was a spiritual leader: Mahatma Gandhi.

Gandhi should properly be regarded as the father of the green movement. Every idea now being pursued within the movement, from the recycling of waste materials to organic farming, was pioneered by Gandhi within the villages of India. He was convinced that, despite its military and economic prowess, Western civilisation is evil because it is at war with Nature: in seeking to dominate Nature for the benefit of humanity, Western capitalism is in truth destroying Nature, raping the earth and polluting the air. With the Indian village as his laboratory, he sought to create a pattern of society in which human beings could live in harmony with their environment, caring for their fellow creatures and cherishing the beauty of the land.

Gandhi was a shrewd and skilful politician, with an enviable gift for persuading people to put his visions into practice. Yet he never tired of reiterating that these visions would crumble to dust unless they were firmly based on spiritual faith. And far from regretting the diversity of religions in India—Hinduism with its many sects, Islam, Sikhism, Christianity and others—he rejoiced that in their different ways they each expressed the same universal spiritual truth. Thus, by each person remaining loyal to his or her own religion, human civilisation could be reformed and harmony with Nature could be restored.

Gandhi himself was of course a Hindu and based his own ideas on the Bhagavad Gita, the Hindu scripture in which Krishna describes three ways in which human beings attain unity with God. The first is the

way of knowledge (*jnana-yoga*), in which through contemplation the person perceives the divine unity of
all creatures. The second is the way of devotion (*bhakti-
yoga*) in which the person comes to love all creatures
as divine. And the third is the way of action (*karmayoga*),
in which the person puts his or her knowledge and
love into practice, serving the needs of all creatures.
Gandhi saw these three aspects of faith as a programme
for moral, social and political change. The first task is
to open people's eyes to the glories of Nature, which
in turn will open their hearts in love to all living things.
Then they will automatically create a new society,
in which they live in harmony with one another, and
with the animals and plants with whom they share their
land.

One of Gandhi's closest associates was an Anglican
priest, C.F. Andrews, who was himself imbued with the
theology of F.D. Maurice, the great nineteenth-century
Christian Socialist. Andrews followed Maurice in believing that the doctrine of incarnation must be at the centre
of a Christian social vision: by becoming flesh in the
person of Jesus Christ, the divine *logos* has made all flesh
holy, placing upon us a divine obligation to feed and care
for people's bodies as well as their souls. Gandhi inspired
Andrews to take the doctrine one stage further: to assert
that Jesus Christ represents not just human beings, but
all living creatures down to the smallest insect and flower.
Thus every creature is holy, to be loved and cherished
for its own sake.

It is easy to dismiss such people as Gandhi and

Andrews, and other similar prophets, as starry-eyed mystics whose visions floated way above the heads of the populace. Yet neither of them was asking people to follow some esoteric spiritual path; nor did they imagine that everyone would share their visions. Their purpose was to achieve a subtle but important change in attitude: that humans should regard other species with the same respect as they regard their own. This does not require any special mystical insight, nor any leap of imagination. On the contrary, they said, humans naturally possess this respect; the problem is that Western culture suppresses it. If that suppression could be lifted, then both the politics and the morality of our civilisation would be transformed.

For the past two centuries the Christian churches have seen their mission as converting people to Christianity, and thus swelling their own numbers. But this was not the main mission of Jesus himself, and if our purpose is to spread the gospel of '*logos* made flesh', it should not be our present mission either. Jesus Christ proclaimed 'the kingdom of God, on earth as in heaven', and in his teachings he drew a vivid picture of life within God's kingdom. After the resurrection of Christ the apostles recognised that Jesus himself embodied that kingdom, manifesting God's rule within his own body and spirit. So in going out to preach the gospel, they were proclaiming that in unity with Christ all God's creatures could live in harmony. In Paul's words, God's plan 'is to bring all creation together, in heaven and on earth, with Christ as head'.

In common with every religious group, before and since, Christians soon began to lose sight of their original vision, and concentrated instead on their institutions. They no longer tried to address the wider social and moral issues facing the human race, but focused solely on the well-being of the Church. The Christian mission came to be seen as attracting more people to join the Church, as an end itself. And since the sixteenth century the Western churches have sent evangelists to every corner of the world in order to swell the ranks of believers.

Happily men like Maurice and Andrews—plus many others across the world—have sought to return to the original Christian mission. They have been prompted not simply by their own personal faith, but a passionate conviction that the Christian gospel is more urgently needed now than ever before. Only if people can, in their own fashion, perceive God's *logos* infusing the entire creation can the planet itself be saved. To the dismay and disgust of many of their fellow Christians, this new breed of missionary does not assert that the Christian religion possesses a monopoly on spiritual truth. On the contrary, they rejoice that the truth cannot be confined to any particular religion or institution, but is to be found wherever men and women sincerely seek it. Thus they willingly stand shoulder to shoulder with Hindus, Muslims, Jews and Buddhists, and with any sincere seeker who will share in a common mission, to proclaim the divine unity of all creation.

Jesus saw himself standing firmly with the tradition of the great Hebrew prophets. And this too is where the new missionaries stand. Like the old prophets, they see no gulf between religion and society; instead they regard religion as the inner spirit of society, transforming its outward actions. Thus religious insight should infuse politics and industry as well as family life and personal relationships. This is not to say that they are indifferent to religious institutions. Just as Isaiah believed Israel should be a 'light to the world', so the new missionaries want the various religious groups and bodies to act as beacons lighting the way.

Practical Religion

When Gandhi began his mission in South Africa, he founded a community whose way of life embodied the values he was trying to preach. People of all races and creeds joined together in building simple houses, tilling the soil and sharing meals. And when he returned to his native India he did the same, starting an *ashram* where a small group could learn to live in harmony with one another and with the other creatures of God's earth.

Two thousand years earlier the first Christians had also formed a community in Jerusalem, embodying the teachings of Christ. After the coming of the Holy Spirit at Pentecost, they pooled all their possessions, 'distributing to each according to need'; and they met daily to pray and eat together. We do not know how long that first community lasted; but for the next three centuries Christians were renowned for their close fellowship, sharing their material goods and their food, as well as worshipping together. And when, after Christianity became the official religion of the Roman Empire, this warm corporate spirit began to fade, monasteries and convents sprang up in the desert and in the countryside, imitating the original community in Jerusalem.

During the past two decades there has in the Western countries been a fresh flowering of prophetic communities. As people have come to perceive the profound crisis within Western civilisation, they have begun to experiment with alternative patterns of life. Some have been directly inspired by Gandhi, while others have looked for roots within the Christian tradition, imitating the early Church and the monasteries. Some have no specific religious affiliation, while others are Christian.

As in the early days of monasticism, there have been numerous heroic failures, with communities collapsing amidst broken marriages, lost fortunes and shattered dreams. But failures and successes alike are proving an invaluable test-bed for a future, sustainable civilisation. And since, like the early monasteries, they attract numerous visitors, their influence on our ideas and attitudes is out of all proportion to the numbers directly involved. To tens of thousands of people they are a symbol of hope, a sign of the 'new age' to come.

We need more such communities—communities which embody the universal *logos* in their pattern of life. The present level of interest in them is tiny compared with the symbolic power which they will hold in the next century, as the green movement gathers momentum. And one must hope that the institutional churches will encourage such communities, both by providing land and resources, and by giving their spiritual support. Yet these communities are worthless unless in some way their vision is taken out to the towns and villages where the majority of people live, and given expression in symbols and rituals which all can share.

Gandhi understood this perfectly. There were some among his followers who poured scorn on the popular religiosity of India, regarding it as debased and corrupt. Gandhi, however, urged people to treat all forms of religion with respect as the vehicles of deeper spiritual truth. He recognised that a sacred building in the midst of a community, such as a temple, church or mosque, makes the community as a whole sacred. Similarly sacred places, such as a holy mountain or river, make the whole of Nature sacred. And sacred times, festivals marking the seasons, and also the special moments in people's lives, make time itself sacred. Thus popular religion in all its varied expressions is the means through which we perceive the divine spirit in life as a whole.

In the West today one might imagine that there is very little popular religion left, judging by attendance at church on Sunday mornings. But religious observance has never been a good measure of religious sentiment. Despite the decline of the institutional churches, there is an astonishing renewal of popular religion. The most obvious sign is the extraordinary reverence now shown to ancient sacred buildings. As any priest will ruefully admit, even if his congregation has dwindled to a beleaguered handful, there will be no shortage of people willing to raise money for the church and to decorate it with flowers. Throughout the Western world churches and cathedrals are more lovingly maintained now than ever before—and their sublime beauty speaks more eloquently and to a far wider group of people of the presence of God in our midst than a thousand sermons could do.

Less obviously religious, but of even greater spiritual importance, are our sacred places. Instead of sacred mountains and rivers, we have nature reserves and national parks, where fauna and flora can flourish undisturbed. And although politicians may crassly justify expenditure on these areas as investment in tourism, the love and reverence in which we hold them arises from a spiritual conviction that animals and plants should be conserved for their own sakes. Even if a bird sanctuary or undrained marsh is quite inaccessible, most people regard it as precious; and when it is proposed that such places are disrupted or destroyed for commercial gain, the public outcry is deafening. The cynic may retort that these reserves and sanctuaries do little to protect the species of the world from man's depredations. But their significance is spiritual, not practical, expressing our inner commitment to all living creatures.

It is astonishing, in our busy, fevered lives, how tenaciously we cling on to our sacred times. Harvest festivals and remembrance rituals remain popular. And, even amidst the commercial clamour, Christmas remains a religious celebration of profound importance for many people. Baptisms, church weddings and funerals are also cherished by many who could barely utter a word of the Christian creed.

These popular forms of religion are easily dismissed as spiritual indulgence. Even the clergy are inclined to disparage them as 'folk religion'. Yet their popularity is their strength. They are the means through which ordinary people come to experience themselves as part of a global spiritual community. They are thus the true spiritual source of the green movement. Gandhi realised earlier this century that his great crusade would quickly run into the sand if it appealed only to people's political interests. If it was to be a popular movement that could survive setbacks and disappointments, it had to tap people's spiritual energy—hence his respect for popular religion. Similarly if the green movement is to mobilise popular opinion, and also to persuade people to embrace the enormous changes in personal life-style which the green transformation will require, it too must harness the religious emotions of ordinary people.

And, as Gandhi realised, popular religion transcends human barriers, both of religion itself and of politics. Bishops and theologians may be acutely aware of the differences between one religion and another, and may assert the uniqueness and superiority of their own faith. But to those ignorant of the minutiae of theology, the essential spirit of each religion seems remarkably similar. Thus Gandhi urged people both to remain loyal to their own faith, and at the same time to acknowledge their unity with people of other faiths in the common spirit.

Religion is about symbols. The *logos* communities symbolise the values that society as a whole must adopt if the human race is to survive and flourish in the third millennium. The sacred places, buildings and times to which people cling tenaciously symbolise those same values, to be found in every human heart and in every human community. Symbols are powerful. Far from being a luxury in which we can indulge after the planet has been saved, religion alone can provide the spiritual power and moral courage to enable us to make the changes that are needed. It is not some selfish form of escapism to live in a community, nor is it some idle fancy to maintain ancient churches, to celebrate ancient festivals, and to preserve sites of natural beauty. The green movement rejects religion at its peril; for if it harnesses the power of religious symbols, it will win the heartfelt support of the general public. It is only by encouraging and enhancing our own sense of the sacred that we shall find the motivation and the wisdom to treat the whole world as sacred.